More prais

"*Koala* is a winner in many diff[...] [...] of so many other animals, warn[...] [...] irreversibly destroying our magnificent planet and the only way forward to save koalas, other animals including humans, and all of our homes, is to strive for peaceful coexistence grounded in compassion, respect, and kindness. Easily readable, with a welcomed personal touch, I highly recommend this book."

—Marc Bekoff, author of *A Dog's World*

"Leaving no stone unturned, *Koala* makes great strides to advance our knowledge of this largely misunderstood animal."

—*BookPage*

"An impassioned and informed plea for the conservation of Australia's flora, fauna, and wild places. This is natural history and science writing at its best."

—Peter Menkhorst, *Australian Book Review*

"Charming and intelligent. . . . A vivid journey into a fascinating corner of the natural world."

—*Kirkus Reviews*

"Entertaining. . . . [An] insightful peek into the world of koalas. . . . [T]his is the outing animal lovers didn't know they needed."

—*Publishers Weekly*

KOALA

ALSO BY DANIELLE CLODE

Killers in Eden

Continent of Curiosities

As if for a Thousand Years

Voyages to the South Seas: In Search of Terres Australes

Prehistoric Giants: The Megafauna of Australia

A Future in Flames

Prehistoric Marine Life in Australia's Inland Sea

From Dinosaurs to Diprotodons: Australia's Amazing Fossils

The Wasp and the Orchid

The First Wave (with Gillian Dooley)

In Search of the Woman Who Sailed the World

John Long: Fossil Hunter

KOALA

The Extraordinary Life
of an Enigmatic Animal

Danielle Clode

W. W. NORTON & COMPANY
Independent Publishers Since 1923

For information about permission to reproduce selections from this book, write to
Permissions, W. W. Norton & Company, Inc., 500 Fifth Avenue, New York, NY 10110

For information about special discounts for bulk purchases, please contact
W. W. Norton Special Sales at specialsales@wwnorton.com or 800-233-4830

Manufacturing by Lakeside Book Company
Production manager: Louise Mattarelliano

The Library of Congress has catalogued the hardcover edition of this book as follows:

Names: Clode, Danielle, author.
Title: Koala : a natural history and an uncertain future / Danielle Clode.
Description: First American edition. | New York, NY : W. W. Norton & Company, Inc.,
2023. | "First published in Australia in 2022 as Koala: A Life in Trees by Black Inc., an
imprint of Schwartz Books Pty Ltd." | Includes bibliographical references and index.
Identifiers: LCCN 2022030777 | ISBN 9781324036838 (hardback) |
ISBN 9781324036845 (epub)
Subjects: LCSH: Koala—Australia. | Koala—Australia—Ecology. |
Koala—Australia—Habitat.
Classification: LCC QL737.M384 C56 2023 | DDC 599.2/5—dc23/eng/20220705
LC record available at https://lccn.loc.gov/2022030777

ISBN 978-1-324-07449-6 pbk.

W. W. Norton & Company, Inc., 500 Fifth Avenue, New York, N.Y. 10110
www.wwnorton.com

W. W. Norton & Company Ltd., 15 Carlisle Street, London W1D 3BS

1 2 3 4 5 6 7 8 9 0

For all the environmentalists, carers, conservationists, researchers, rangers and nature lovers who work so hard to protect the future of our wildlife

The author acknowledges the Traditional Owners of Country throughout Australia and recognises their knowledge and continuing connection to the land, waters and wildlife. She pays her respects to Elders past, present and emerging.

CONTENTS

Fossils of koalas have been found across the southern and eastern regions of Australia at the sites named on this map. Over the last two hundred years, the distribution and abundance of koalas has retracted east. They are now least abundant in the palest shaded areas and most abundant in the darkest and have re-established wild populations in their former range in South Australia.

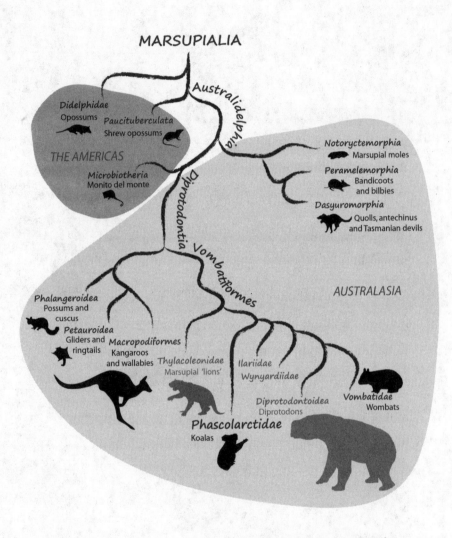

The koala family tree, with common names and illustrations of representative members. Extinct groups are in grey.

I

INTO THE WOODS

A cool breeze ruffled the koala's fur, causing her to stir in her sleep. With her head tucked into her chest, she was barely visible in the waning light – a grey mass wedged in the fork of a tree. Her large ears rotated slowly, scanning the surroundings. She lifted her head, eyes still closed, breathed in the dampening night air and let the sounds and smells of evening wash over her.

She could hear the creek burbling below and the rhythmic nightly chorus of crickets. The frenetic noise of the day had subsided. Most of the birds had already retired to their roosts, save a lone gang-gang cockatoo, its low creaking cry echoing across the valley as it flew past in search of its family.

Sleep beckoned. It was too soon to feed. The trees still hummed from the energy of the departed sun. It would be some hours before they drifted into their nightly cycle of respiration, when their defences dwindled and their leaves were at their most succulent and tasty.

Even so, it was time to move. The trees here were redolent with the scent of other koalas and their leaves already heavily browsed.

She stretched a leg and scratched behind her ear before moving down the trunk of the tree with sudden swiftness. She dropped with a crunch onto the shedded piles of dried bark and headed off on a path she had not taken before.

Each night took her on a new route. Each trail was unfamiliar, filled with hazards and perils. Sometimes her course took her through patches of heath, over rocks, across creeks and clearings or into pockets of forests. But every night, long before dawn, she'd stop and sit on her haunches, then return to the trees she had left.

These treks were not long but they were tiring and did not give her much time to eat when she returned. Sometimes she was forced to eat during the day, when the leaves were sharp and bitter. But she had no choice – she had to continue.

Tonight, the moon was setting, a tiny sliver of silver drifting down towards the horizon. She soon left the comforting scents of the patch of trees where she had been born and spent most of her life, and headed out across an open clearing. It was easy travelling at least, even if she was vulnerable to the silent swoop of powerful owls. Clouds scudded across the sky, concealing even the bright starlight of the Milky Way and providing a brief cover of darkness.

The vegetation thickened into heath, and she wandered between the bushes, weaving an unsteady path until the heath gave way to low scrub. She quickened her pace through the open undergrowth. Smelling damp ground, she veered off towards the promising scent of water.

Before long, the dampness condensed into a trickle and then a creekline. She followed the trail down into a gully. Trees rose overhead and ferns obscured her path, but she persisted, splashing through intermittent pools of water.

Her stomach grumbled and her feet ached. She'd gone further than she'd managed on any previous night. She needed to find food or return, but something in the air lured her on just a little bit further.

A rocky outcrop blocked her path, a precipice opening up before her. She turned and dropped backwards over the edge, her feet scrambling for a foothold, her large black claws splayed like grappling hooks over the smooth surface. Reaching the bottom, she stopped and sniffed. The gully was damp and cool and the trees were large. The blend was just right. She could smell the crisp wattles and the sweet bursarias, the spicy undertones of the brackens, but beneath it all lay the smooth rolling perfume of a mature manna gum. She headed towards it, reeled in by the intoxicating scent.

Bark and leaves crunched beneath her feet as she approached the old tree, rising vast and broad above her. She stretched up on her hind legs, her forearms wide against the tree, taking short, deliberate breaths. Something was missing.

She breathed in again, sucking the air up through the roof of her mouth. Not a hint, not a trace of the smell she had always lived with. The smell of other koalas.

This patch was hers and hers alone. She bounded up the tree, into the grey-green leafy curtains of abundance. She'd found her new home. ✐

1

Koalas Rare and Plenty

Late one night, as I am driving home from a meeting in town, a koala crosses my path. Its furry white rump sashays casually in the sweep of my headlights as I round a bend in the road. I slow the car to a halt and lower my high beam, waiting for it to cross. The koala pauses, sits down in the middle of the road and looks over its shoulder, as if glaring indignantly at my impertinence.

I check my rear-vision mirror for approaching traffic. It's a quiet area but I don't want to be sitting on a blind corner on an 80-kilometre-an-hour rural road. And I am impatient to get home.

I inch the car forwards, hoping the koala will move, but it looks away and doesn't budge. I can tell from the angle of the fluff on its ears that it is listening carefully. I wonder if my headlights are still too dazzling, so I put on my hazard lights, blinking orange, and dim the headlights further. Koalas are not prone to making rash decisions. They are not to be rushed. I take a deep breath and prepare to wait.

In the dimmer light the koala seems more comfortable. It eventually rises onto all fours and, without a backward glance, saunters off into the roadside vegetation, unhurried and apparently unperturbed.

I drive the rest of the way home more slowly. This stretch of road is full of wildlife. A mob of grazing kangaroos startles at my approach, lifting their heads and twitching their ears, each of them tensing like a tightly coiled jack-in-the-box ready to spring away from, or into, danger. Southern brown bandicoots, lizards, snakes and even echidnas are also occasional casualties of collisions here. I don't need to get home in such a hurry. Perhaps I should be grateful for this unexpected interaction, rather than irritated by it. I could probably do with slowing down the pace of life and taking my time – being more like a koala.

Encounters with koalas are not rare where I live, in the Adelaide Hills. They are nowhere near as abundant as the kangaroos that scatter across the paddocks every nightfall, but most people who live here have had a close encounter with a koala at some point, and have a story to tell about them. They wander across the lawn in front of my mother's lounge-room windows, pausing to inspect her as she watches TV. They bail up and are bailed up by dogs in backyards and in bushland. In spring their grunting bellows reverberate through the forests, and in summer they come down to drink water from ponds and pools.

The Adelaide Hills are part of the Mount Lofty Ranges and run along the length of one of Australia's smaller state capitals. Modest farm-holdings are interspersed with growing commuter townships and the swathes of remnant forest protected in conservation parks or reserved for forestry and water catchment. It's an area rich in biodiversity and wildlife, but it has not always been home to koalas. When my parents grew up in Adelaide, they never saw koalas in the wild. Instead, they visited the 'koala farm' in the city parklands to see a regular crop of young joeys, alongside high-diving seals, buckjumping pigs, pythons and kangaroos. After the koala farm closed, people travelled to Cleland Wildlife Park, in the hills overlooking Adelaide, to watch, photograph and hold koalas.

I ask a neighbour who grew up in this area of the hills whether she remembers koalas here in her childhood. She shakes her head.

'I spent most of my time riding my bike and exploring the bush-land,' she says, 'but there were no koalas here fifty years ago.'

'We often went blackberrying through the Adelaide Hills in the 1940s and 1950s,' another friend adds, 'but we never thought that we could see koalas in the trees.'

Koalas are largely native to the eastern mainland states. Their natural range stretches from north Queensland across New South Wales and down to Victoria, with just a small pocket jutting into the far south-eastern corner of South Australia. At one time or another, I have lived in and around the forests of Queensland, New South Wales and Victoria, but I never saw koalas there in anywhere near the abundance I have seen them here. Friends who live on the eastern coastline tell me that sightings in the wild are rare there. Most of them have only ever seen koalas in zoos and parks.

I don't know when exactly koalas arrived in South Australia, or whether it was accidental or intentional, or why they have expanded and spread so much since the 1980s. But by 2019 there were estimated to be around 150,000 koalas living in the Adelaide Hills and Mount Lofty Ranges, a further 48,000 living on Kangaroo Island, as well as smaller populations in the south-east corner of the state, on the tip of Eyre Peninsula and along the River Murray. In places, they are so abundant that contraception is now a regular part of their ongoing management, due to over-browsing of trees.

Populations elsewhere are not doing quite so well. If the estimates are accurate, there are now more koalas living in the south coast states of South Australia and Victoria than the total number said to remain in the east coast states of New South Wales and Queensland – whose combined koala populations are reported to have dropped below 80,000 individuals.

I keep seeing alarming reports of koalas being described as 'functionally extinct' in the wild – a term usually used for species with only a single surviving individual. Images flash on TV of homeless koalas perched on the fallen trees of a logging coupe, or bandaged and burnt after bushfires, or hit by cars or injured by dogs from ever-increasing urban development. Koalas have been steadily declining in New South Wales and Queensland for decades. In 1988, the New South Wales government supported a 'Koala Summit' to bring together researchers, wildlife managers, land-use planners, private and public interests, as well as state, local and federal governments, to help plan for the koala's future. Thirty years later and nothing seems to have made any difference. I wonder what is happening with this most iconic of Australian animals. What makes them do so well in some places and so poorly in others? What do koalas need to maintain a healthy and robust population in all parts of their range? What are we doing wrong?

I realise that I just don't know enough about this famous Australian animal to answer these questions. Just how many koalas are there? What habitat do they need? What exactly do they eat in different parts of the country?

It amazes me that a creature this iconic and distinctive to Australia is so mysterious. When Europeans first arrived here, it took them a decade to even notice these strange animals in the trees above, and it was thirty years before they were scientifically described. Two centuries later, scientific research into this remarkable animal has gathered pace only in the last fifty years.

The more I try to find out about koalas, the more I realise there is to learn. I'm impatient to solve the mysteries surrounding them, to unpack the myths, inaccuracies and fabrications. Primates are noto-riously impulsive, impetuous and impassioned – the product of our 'monkey brains' – and as unlike koalas as a tortoise to a hare. I search the library shelves for answers. There are hundreds of children's story-books about koalas, and yet barely a handful of general interest books about them for adults.

'Perhaps there is not that much to know,' says my husband. 'They live in trees, eat gum leaves and sleep a lot. Maybe they are just not very interesting?'

I don't think that's true. There are more scientific papers and gov-ernment reports on koalas than any other Australian animal. There is a lot more to koalas than meets the eye. If I want to really understand koalas, I'm going to have to slow down and see things at their pace and from their perspective. It's going to take some time.

Thirty years ago, when I was a university student studying animal behaviour, I lived in a share house in the hills. I don't remember any koalas in the trees over our house, although I think one might have visited a neighbour once. But I do remember the yellow warning signs emblazoned with a koala silhouette on the roads winding up into the hills. One night, cars in both directions stopped as we waited for a koala to cross the narrow road. Caught in the headlights, it struggled to get over the slippery steel barrier on the steep roadside. The driver of our car leapt out and carefully gave the koala a lift from behind, as it scrambled up the barrier and over into the safety of the bush beyond. I wondered then how many koalas had not been so fortunate, and why the barriers were not made with more clearance or with better grip to make the road safer for wildlife as well as people.

But other than the occasional sighting of a furry grey blob in a tree, my first close encounter with a koala came when I worked as a casual zookeeper at the Adelaide Zoo, during a 'gap year' between my undergraduate degree and going overseas to study conservation biology. My job was to fill in for other keepers if they were unwell or on holiday or on parental leave. That role took me across many different rounds, working with a wide range of birds and mammals – from the fennec foxes of Saharan Africa to the colourful macaws of the South American jungles.

The Australian animals were the zoo's centrepiece. The first thing people saw when they arrived was a large rocky cairn, attractively draped with a colony of exquisite yellow-footed rock wallabies, whose rescue from imminent extinction through captive breeding, habitat protection and reintroduction has become one of the zoo's great success stories.

I learnt how to look after echidnas, who trundled busily around their enclosures like small, spiky tanks, ignoring the noisy prattle of fruit bats roosting above them. I prepared seed mix and chopped vegetables for the secretive stick-nest rats; flowering gum leaves, chopped fruit and a bit of boiled egg for the brush-tailed possums; and mealworms, mice and fly pupae for the feisty little kowari. I boiled carrots for the cassowaries, fed fish to the little fairy penguins and weighed tins of specially mixed seed for the parrots. I cleaned, swept, raked and shovelled, carted and carried, rinsed and scrubbed. If there was extra time, we worked on strategies to distract the hyperactive Tasmanian devils from pacing, or flooded the magpie goose enclosure to encourage nesting.

The koalas, though, were one of the easiest animals to care for. Always amenable to being displayed, they would happily sit in the freshly cut gum branches brought into their enclosure each day to adorn the stags or trunks of large old trees that formed the structural framework of their exhibit. They never trashed their enclosures, like the destructive red-tailed black cockatoos, or disturbed prudish visitors with antisocial displays of promiscuity or violence, like the monkeys and baboons. They were easy to clean up after. Only the gardeners complained about their insatiable demands for a variety of fresh gum leaves. And they were entirely unperturbed by the steady stream of adoring admirers parading past them.

There were almost never any vacancies on the koala round. They had a steady stream of adoring keepers too.

Not everyone loves koalas, though. The internet is full of stories about how koalas are nowhere near as cute and cuddly as people think – that they are aggressive, that they smell, that they wee on you, that they are feral pests destroying forests. They are described as slow, torpid, simple, small-brained, primitive, degenerate, overspecialised, inbred, disease-ridden and maladapted. In the past, we have hunted them to near obliteration, and every day we continue to ruthlessly cut down their trees, clear their forests, fragment their habitat with roads and fences, and replace their homes with our own. Maybe that's easier to do if you think they are doomed to extinction anyway?

The internet may not always be the best place for accurate information, but the conflicting perspectives on koalas reveal a broader ambiguity in our attitudes towards wildlife. Koalas can be both affectionate and aloof, beloved and belligerent. They are neither bears nor soft toys. They are simultaneously threatened with extinction in some regions yet dying from overpopulation in others. They are vulnerable to climate change, deforestation, disease, habitat fragmentation, bushfires, overdevelopment, forestry, dogs and cars. They are both a symbol of Australia's unique wildlife and emblematic of the devastation we have wrought on so many species.

In 1938, wildlife biologist Ellis Troughton was certain about what the future held for koalas.

'No one can doubt that ultimate extinction awaits the gentle and highly specialized leaf-eating koala,' he declared.

Koalas face many challenges, but I wonder if we are underestimating the achievements and resilience of this species. They are supremely well adapted to their environment. The koala is one of the few species that has managed to exploit the foliage of the eucalypt trees which dominate Australia's modern forests. Unlike the thirty-four mammal species that have become extinct in Australia since European colonisation, koalas have not only survived, but expanded and thrived in parts of Victoria and South Australia.

They may well be one of the most loved and best-known icons of Australian wildlife, but they deserve better than to be simultaneously patronised as fluffy children's toys and left to suffer from increasing diseases in ever-reducing habitats. Surely they warrant our care and attention, for us to focus on what they are, how they live and what they need – who they are as animals?

There is so much more to know about koalas than just a sweet face on a postcard – their history, evolution, biology, ecology, their interactions with humans and other predators, where they have come from and what their future holds. This book is their story.

II

FROM FOSSILS
AND BONES

The koala dropped to the ground. The sun was rising, so it was time to retreat to a perch in a tall tree a short distance away. Following a wombat trail, the koala paused when it reached the edge of the swamp. It hesitated, listening to the sounds of the forest as the morning sun stretched long tendrils of warmth through the canopy. Here on the forest floor, the world was cool and damp. Insects whirred across the still surface of the water as the pulsing trills of the painted frogs gradually subsided into slumber. A potoroo scuttled through the undergrowth, startling a small mob of wallabies from their grazing. They sat upright on their haunches, eyes wide and ears twitching. But there was no cause for their alarm. No thylacines prowled nearby; no thylacoleos crouched in the nearby trees.

The koala looked again at the swamp. There were threats here other than marsupial carnivores. These waters were home to grazing hippo-like zygoma-turines, whose sheer size made them dangerous. Palorchestes, with long bark-ripping claws and powerful forearms, were also best avoided. But the real danger was less obvious. Wonambi, 7-metre-long constrictors, lurked near the waterways for unwary victims.

A turtle lifted its head from the water, before submerging with a loud plop. The koala grumbled and turned away, keeping to higher ground as it made its way around the sedges and reeds that bordered the swamp.

The woodland quickly opened out further from the water, patches of grassland encroaching into sunlit openings left by taller trees that had aged, senesced and fallen. Reaching the base of an old river red gum, the koala paused again, a sharp scent assailing its nostrils. Something else had recently climbed this tree. There were not many creatures that would deter the koala from its favoured roost – neither gliders nor giant ringtail possums, not even the carnivorous, quoll-like glaucodon. But this new resident was different. Even as the koala sniffed again, it heard a low throaty rumble from above. The koala scampered away with uncharacteristic speed. There was no point in challenging this one. The new resident was an entirely different species of koala – far bigger and more robust, literally twice the height and heft. A giant among koalas.

A young manna gum beckoned nearby. It was a poor substitute for the big red gum. The displaced koala scaled it quickly, content to find it unoccupied, at least by any others of its kind, and settled into a convenient fork to sleep, its nostrils detecting only the cool reassuring aroma of eucalyptus. ☙

2

Dropbears in the Family

The visiting journalist glances at the camera, clearly anxious about her current assignment. She's in a typical Australian suburban backyard. A wide yellowing lawn is adorned with a half-full rotary clothesline, a handful of kids' toys and an industrial-sized above-ground swimming pool. It's hard to imagine anything remotely dangerous here.

'How're you feeling?' asks her Australian guide, as he buckles her into some kind of heavily padded tactical armour that he describes as a 'dropbear suit'.

'I'm a bit worried,' she replies, 'about why I need this level of protection.'

'Dropbears are a close cousin of the koala but they are actually really vicious, sort of like a dingo and a domestic dog,' her companion explains. 'They're bigger, they've got longer claws. They've actually got really small fangs, and the interesting thing is that they have a really mild venom. It's not like a snake venom that can make you sick, but it just causes a lot of local irritation. It's actually the third most common injury that we see in tourists in Australia.'

The reporter looks terrified as they buckle her into boots, leather gloves and protective eye gear. She's from Scotland, where the fiercest creature you might encounter is the odd wildcat.

'No fast moves,' one of her colleagues explains. 'Keep calm – dropbears can sense if you're worried.'

They pass her a large grey koala, who grips her shoulder tightly as he gazes placidly at the overhanging vegetation, probably thinking it looks like a better place to be.

'What's it doing right now?' the reporter asks, glimpsing anxiously at the distracted koala. Off camera, her colleagues erupt in panic.

'Quick!'

'Take it off her!'

'Get the dart gun!'

As the koala is retrieved, the tension suddenly shifts. Smiles and laughter break out among the crew, and the reporter curses in mock fury, realising she's just been had by a classic Australian joke – to convince overseas visitors that one of the cutest and gentlest creatures is actually our most deadly predator.

Australia has a reputation for dangerous animals. We do have more than our fair share of venomous snakes, spiders, jellyfish and shells, but they rarely kill anyone. Sharks and crocodiles will occasionally attack unlucky swimmers. Kangaroos can be aggressive if cornered, and even platypus have venomous spurs, so why not a koala? But anyone who has met an amiable, leaf-munching koala will know that violence is, for the most part, the furthest thing from their minds. The only known victim of a koala is the dignity of gullible tourists. And the greatest threat from being handed a koala is probably an ill-timed piddle.

The origin of the dropbear myth is a bit of a mystery. Some people have suggested it is a primeval memory of a prehistoric marsupial lion, *Thylacoleo carnifex*, that lurked in ancient trees to ambush the unwary. But the joke seems to have gained most popularity in the

1980s, after Paul Hogan faced 'killer koalas' in a precursor TV skit to his famous Crocodile Dundee character. Even so, it is fair to say that if a giant or carnivorous koala ever did exist, the one place we might find it would be in the distant past.

In 1889, one of the first curators of the Queensland Museum, Charles W. de Vis, found an intriguing fossilised shinbone in the collection. De Vis knew, from other fossil species, that Queensland had once been more widely covered in 'vegetation of tropical luxuriance'. While the forests on Australia's east coast today, and the animals they support, are largely restricted to the coastal fringe east of the Great Dividing Range, in the past this forested area spread further west into the now arid inland plains. So, de Vis was surprised by the lack of fossils of tree-dwelling animals, like possums and, more particularly, koalas – evidence for which he had long 'waited and searched in vain'.

The shinbone de Vis found had features that suggested it was from a close relative of the koala – but one that was much, much bigger. He also spotted a skull fragment which he thought might belong to the same species. It closely resembled modern koalas except for an unusual canine tooth or 'fang'. De Vis was thrilled to find the first evidence of a koala in the fossil record, so he named this new species *Koalemus* – humorously, but rather unkindly, after the little-known Greek god of stupidity.

Had de Vis found a giant fanged koala? Was this a real-life precursor to the mythical dropbear of contemporary Australian folk stories? Probably not. There are plenty of strange, fanged beasts in Australia's fossil history – including herbivores that had adapted to eat meat. The koala's fearsome cousins – the marsupial lions – were highly specialised for carnivory. Fang-like teeth and well-developed canines, however, do not always indicate meat-eating. There are many animals that have fangs for reasons other than food – pigs, primates, deer and even tiny fungi-loving bandicoots.

The identity of *Koalemus* would remain a mystery for decades to come.

Koalas are singular creatures: idiosyncratic and inimitable. They are sometimes described as being 'like bears', 'like wombats', 'like sloths' or 'like pandas'. They share some parallels, some traits with these creatures, but they are not in any way 'like' them. Koalas are simply unlike anything else we know of.

Even so, koalas do not exist in isolation. They are an integral part of a complex and colourful ecosystem of creatures adapted, like them, to arboreal life and to Australia's capricious climate. And they share their past with a characterful array of cousins. They have both a neighbourhood and a family history.

'And who is their family?'

The censorious tone of an elderly great-aunt grates in my memory. This was always a question bound to aggravate rebellious nieces, but in this case it is useful. Koalas have a place, in the now and in the past. I need to work backwards like a genealogist, to construct a koala pedigree, a family tree that shows me where everyone fits. If I want to understand koalas, I need to know where they have come from.

Like most Australian mammals, koalas are marsupials – as are huge hopping kangaroos, miniature gliders, ant-eating numbats and beaming quokkas. Rather than growing their babies in swelling bellies through pregnancy, marsupials give birth to tiny hairless young barely the size of a jellybean. These neonates crawl through their mother's thick fur into a soft, sterile pouch, lock themselves onto a nipple and mature there until they are ready to emerge through a 'second birth' from their mother's pouch into the world.

Australasia is the heartland of marsupial evolution. The furry pouches in which marsupials keep their infants are the outward sign of a reproductive system that immediately separates them from the much more common eutherian mammals (formerly, although inaccurately, known as 'placental' mammals). Of the over 300 marsupial species living today, more than two-thirds are native to the once-connected landmasses of Australia, New Guinea and their associated islands. And what's more, almost two-thirds of Australia's native mammals are marsupials. No other landmass in the world is as dominated by marsupials as Australia.

Even so, marsupials are found elsewhere. They are also native to the Americas –which tells us that marsupials evolved and radiated into many species at a time when Australia, Antarctica and South America formed the great southern supercontinent of Gondwana.

I know there is some kind of opossum in North America, although I don't know much about it. An old friend visiting from Louisiana was less than enthusiastic about them. When we lived in Victoria, she was delighted by the kangaroos on our lawn and seeing a koala in the wild was on her wishlist. But when I offered to show her our possum nest boxes, she screwed up her face in disgust.

'Possums are ugly,' she announced.

'How can you say that?' I asked, astonished. 'Possums are lovely – soft fur, big eyes and long tails.'

'We have opossums at home – they're just giant rats,' she declared, shuddering.

'Come and see these ones,' I said. 'Then you can tell me if you think they look like giant rats. We'll have to be quiet.'

We carried the ladder to one of the nearby gum trees and rested it beneath the small wooden nest box. Our nest boxes were in demand, and often on high rotation, variously occupied by a range of lorikeets, rosellas, nightjars, possums, gliders and phascogales. For once, I hoped that the brush-tailed phascogale was not in residence. These rare marsupial carnivores, with their spectacular plumed tail, are one of my favourites but, I have to confess, they could look a bit ratty.

I watched as my friend climbed the ladder, lifted the lid of the nest box and peered inside. After a few minutes she climbed back down.

'Okay,' she said. 'That was cute.'

The box was crammed full of sugar gliders, their tiny striped faces with big eyes blinking, their gliding membranes wrapped between them like a soft fluffy quilt.

'Opossums are nothing like that,' my friend admitted.

It's really not fair to judge marsupials on the only member of the family to have expanded into the northern hemisphere without human assistance. But many people do – even scientists. I suspect the Virginia opossum is responsible for many of the generalisations and assumptions about marsupials. It hardly even does justice to the great diversity of opossums living in the rainforests of Central and South America. I had no idea there were so many: water opossums, pygmy opossums and mouse opossums. There are opossums slender and gracile; species with thick tails, bushy tails, bare tails or fat tails. Others are woolly, striped, coloured, white-bellied and four-eyed (so named for their 'spectacled' markings). But they are all very much opossums: most ranging from mouse- to cat-sized, typically arboreal, opportunistic omnivores.

By contrast, Australian marsupials are extraordinarily diverse in shape, size and ecology. They vary from the minuscule to the massive. The long-tailed planigale is a tiny but rapacious marsupial predator that weighs just 4 grams, one of the smallest mammals on earth at barely a quarter the size of a house mouse. Living in the black soil plains of northern Australia, their flattened and elongated skulls allow them to squeeze between the cracking clay to hunt spiders, centipedes, lizards and even mammals larger than themselves. At the other end of the scale stands the red and grey kangaroos, taller than a human male, with their distinctive bipedal stance, some weighing up to 90 kilograms. From rainforests to arid deserts, from underground burrows to tree-tops, marsupials in Australia occupy all the ecological niches – grazing, browsing, hunting and scavenging, united only by the underlying biology that identifies their common ancestry. And the specialist leaf-eating, tree-dwelling niche occupied by sloths, lemurs and monkeys on other continents is filled in Australia by the koalas and possums.

The Australian marsupial family tree splits into two distinct groups (as shown at the front of the book). One side includes bilbies and bandicoots and marsupial carnivores like quolls and the Tasmanian devil. It also contains some oddities – the stripy little ant-eating num-bat and the sand-swimming marsupial mole. On the other side is a much larger and perhaps more familiar group which includes all the many kangaroos, potoroos, possums and gliders.

From here, the seventy or so species of possums, gliders and cuscuses split away, along with another sixty or so species of kanga-roos, wallabies and potoroos – the bulk of Australia's most visible and abundant species. And we are left with a family that is few in number, much less common and less likely to be seen. These are the Vombatiformes, a group that today comprises three wombats busy burrowing underground and, sitting all by itself high up in the branches, the solitary koala.

Wombat burrows are incredible architectural constructions. Up to 30 metres long and several metres deep, they are vast underground labyrinths built by multiple generations of wombats, the latest of which sometimes guard their ancestral homes with bad-tempered aggression.

In 2018 a group of third-year zoology students from the University of Melbourne set up a camera to record overnight activity at an entry point of a common (or bare-nosed) wombat warren. A wombat duly emerged from its burrow for a minute or two, then returned inside. A couple of minutes later, another animal appeared. It was not a wombat, though. This time, a koala emerged from the same burrow, sniffed the ground where the wombat had just been and then ambled off.

It was an observation that surprised even the students' supervisor, Dr Kath Handasyde, who has been studying koalas in the wild for over forty years.

'Camera traps are amazing tools that allow us to peek into the lives of shy nocturnal marsupials,' she explains. 'Sometimes they reveal things we never dreamed would happen.'

As to what the koala, an almost entirely arboreal animal, was doing down a wombat burrow is anyone's guess. But Kath doesn't think the koala and the wombat crossed paths.

'Bare-nosed wombats can be a bit cantankerous. If the koala had encountered the wombat, it would probably have been chased out and wouldn't have been anywhere near as relaxed as we can see in the photos,' she said.

As their closest living relative, koalas are often connected to wombats. Early European accounts sometimes described them as 'koala wombats'. The very name Vombatiformes means 'wombat-like', as if koalas were some kind of wombat that crawled out of a burrow and took residence up a tree. It's regularly assumed that koalas must have evolved from wombat-like ancestors. 'From the (under) ground up', as one book chapter on koala evolution is titled. I guess they are both short, stocky animals without tails, for all that one climbs up trees and the other digs underground. But in truth there is not much that is wombat-like about a koala, and there is no reason to think that koalas have ever been ground-dwellers.

The differences are much clearer if we look beyond the living species to the now extinct members of this group. Vombatiformes once existed in a great range of shapes and sizes, from 3 kilograms to almost 3 tonnes: creatures that lived in trees, underground and even in swamps and waterways. And they included some of our most remarkable megafauna – Australia's version of the northern hemisphere's Ice Age mammoths, giant sloths, cave bears and sabre-toothed tigers.

Diprotodons were the largest megafauna – rhinoceros-sized marsupial herbivores, pouches swollen with oversized joeys, browsing among the sweeping inland forests that once covered much of Australia. Palorchestes stripped bark from trees with their scimitar claws and tapir-like trunk, while zygomaturus lurked, bunyip-like, in swamps. Others turned to meat for food. Thylacoleo was the size of a leopard and armed with retractable claws in its thumbs, slicing blades for teeth and a bone-crushing bite. Species appeared and disappeared over time, some surviving into human history and captured in cave paintings and legends. Others left little more than broken bones, claw marks raked into a cave wall, or a long trail of footprints sunk into the soft surface of a dry lake bed.

All these strange and diverse species lie between modern wombats and their more distant cousins, the koalas, who sit a little further out, on a side branch of their own on the evolutionary tree. Koalas are as different from wombats – in behaviour, appearance and anatomy – as they are from giant diprotodons or palorchestes.

The one thing that does unite Vombatiformes, though, is their scarcity. Of this once abundant and diverse group, only the koala and a few wombat species survive. Prehistoric wombats once grew to the size of a grizzly bear and numbered several very different species. Today, there are only a few hundred of the highly endangered northern hairy-nosed wombat in the wild. The 'near threatened' southern hairy-nosed wombat is restricted to semi-arid areas of the Nullarbor Plain and isolated pockets in South Australia. Even the common or bare-nosed wombat is decreasing in range. Until recently, they were one of the few Australian native animals that could still be hunted for sport, and their sturdy bodies litter the verges of alpine roads, victims of the fast pace of human life.

And as for the koalas, there is only one modern representative remaining of the twenty or so that have existed at different times through the millennia – *Phascolarctos cinereus*, or 'ash-coloured pouched bear'. While the modern koala once shared its world with one or two koala species, it is now the last living member of its lineage: 'attractive but phylogenetically sterile', in Troughton's words.

I'm not sure if humans are as attractive as koalas, but we are also phylogenetically sterile – that is, lacking in close relatives. Like koalas, our *Homo sapiens* ancestors once shared the planet with other human species: *Homo neanderthalensis*, *Homo naledi* and *Homo erectus* to

name a few. Our nearest relative, the Neanderthals, vanished about
the same time as the last remaining close relative of the modern koala
disappeared. But the koala is more isolated than us within its broader
family. While we can count distant cousins among the other apes
and even monkeys, the wombats are as distantly related to koalas as
humans are to the tiny primate tarsiers that cling to survival, all eyes
and froggy fingers, in the Indonesian rainforests.

Ecologically, as well as evolutionarily, koalas really do sit alone on
their tree.

3
—

The Lakes District

The drive between Australia's two closest state capitals, Melbourne and Adelaide, takes about eight hours. It doesn't leave much time for sightseeing. Every year, as we travelled from one city to the other to visit family, I'd watch as places I wanted to visit zoomed past my window at 100 kilometres an hour.

'Look, there's the giant koala,' I'd say to my children, each time we passed Dadswells Bridge. They'd lift their heads, glance out of the window, then murmur something noncommittal before returning to their screens and books.

The 14-metre-high bronze and steel koala towered by the roadside, marking the gateway to Gariwerd, or the Grampians: a relic of the fashion for 'giant' landmarks in Australian country towns. In the distance, the sandstone outcrops of the Grampians formed their own monumental landmark in the flat horizon. I once camped among these eerie crags as a teenager, and remember the abundance of koalas that sauntered between the tall trees of the campground with nonchalant ease.

'We should stop overnight there some time,' I'd suggest.

'Next time,' my husband would reply, his eyes on the road.

The oldest fossil record of a koala, or Phascolarctidae, is 24 million years old, around the change from the warmer Oligocene age to the cooler era of the Miocene. This period coincides with the rise of many modern mammal and bird groups and the expansion of the grasslands. Recognisable species of dogs, bears, deer, whales and camels appeared, along with owls, ducks, crows and cockatoos. And at the same time as the koalas left the first fossils in Australia, a new group of primates arose in Africa – the apes.

It's possible, though – probable, even – that koalas were around much earlier. Molecular studies suggest that the koalas diverged from the rest of the vombatids around 37 million years ago. But there are no fossils to support this – in fact, there are no fossils of any marsupials in Australia from 26 million to 55 million years ago.

Koala fossils are relatively rare. Perhaps the animals were never abundant. Perhaps they did not inhabit ecosystems that favour fossilisation. But there are fossils of about twenty different koala species, found throughout Australia and spread through time, many known only from a single bone fragment or a tooth. Ironically, many of these fossils were not discovered on the forested eastern coast where koalas live today, but in the now arid inland areas across southern Australia.

It has taken us days to drive to the far north of South Australia from the city, to visit schools and give talks on writing and fossils. Even for Australians, accustomed to lengthy journeys along straight, empty roads, this place is a long way from anywhere. The horizon stretches all around, completely flat with nothing to disrupt the perfect sphere of blue sky above, sliced through by the disc of red earth below.

As I step out of the car, I am surprised to find that the soil is soft and powdery: a fine, undisturbed dust that puffs beneath my feet. The landscape looks lunar, almost Martian. Small rocks lie randomly across the plains, as if sprinkled by a giant hand, and tiny grey plants cling intermittently to life. There is not a tree of any stature to be found for a hundred kilometres or more.

But there were trees here once and the evidence lies scattered around me. Fragments of fossilised wood are on the ground: broken logs that look exactly like wood until you pick them up and feel the weight of stone. Some areas look like fallen forests, others as if the timber had been washed out to sea, then dropped on the ocean floor when ancient seas receded and dried out.

It is not just the traces of trees that speak of a lost world. Reminders of a well-watered past lie in vast, glittering slivers across the flat surfaces. Kati-Thanda, Ngarndamukia, Munda, Pando Penunie and Carle Thulka – also known as Lake Eyre, Torrens, Frome, Hope and Maurice respectively. A few retain their Indigenous names: Lake Bumbunga, Callabonna and Cadibarrawirracanna. These are the last puddles of an enormous, shallow inland sea that stretched across Australia 100 million years ago. As this sea retreated, it was replaced by a multitude of inland saltlakes above an immense underground artesian basin that still lies beneath the desert.

This ancient lake system covers well over 20,000 square kilometres – an area bigger than Israel or Slovenia. Today these lakes have no torrential rivers to supply them with water, and they send no floods to the sea. These are 'endorheic' lakes: drainage basins at the lowest point in a wide, flat land, 15 metres below sea level, with no obvious outlet. They lie dry for decades under a crust of residual salt, filling only once or twice a century when rare downpours sheet water off the inland deserts. The lakes hold this water for days or weeks, as wildflowers bloom in uncharacteristic abundance and desert birdlife sweeps in across vast distances to feed and breed. Animals gather on the edges to feast on the clustered plenitude, until the water eventually seeps underground into the subterranean aquifer or evaporates beneath the unblinking sun.

Over millions of years, these lakes have kept a record of their visitors, written in their shores, of footprints and bodies. Some die on the lake edges, their bones slowly accreting to the surface, gently covered by growing sediments in successive years of flood, rain and blistering drought. These records tell of a once thriving ecosystem of plants and animals in the forests around the lakes.

I turn back to the car. It seems inconceivable that this inhospitable and unlikely terrain, this desert of dried lakes, is the place where the oldest koala fossils, 24 million years old, have been found.

In 1953 an American palaeontologist, Professor R.A. Stirton, arrived in Australia to begin 'a quest for the ancestry of the monotremes and the marsupials'. Stirt, as he was known to his colleagues, brought his wife, Lillian, with him, as well as a graduate student, Dick Tedford. Their destination was a fossil reserve near Lake Callabonna, which had been identified in north-eastern South Australia in 1893. It was not easy to reach, or to work in. There were no proper roads and travel often involved following tracks in the sand or old railway lines. In summer the temperature could exceed 50 degrees Celsius during the day, and on winter nights it could drop to 6 degrees. In the sixty years since the first haul of giant birds and mammals had been found in the area, no-one had been back – and no-one was even entirely sure where the original fossil site was or if it could still be accessed beneath shifting desert sands.

The visiting Americans headed north under the care of the South Australian Museum's taxidermist, Paul Lawson, who was charged with keeping the team and their unreliable World War II truck going. Their first trip was unexpectedly cold, wet and arduous.

'Paul was invaluable,' Stirton wrote in his field notes. 'He never gives up.'

They managed to locate the old fossil site and bring back new examples of the fossils previously found. But what they really wanted were older specimens, which would reveal new information about the evolution of Australia's distinctive marsupials.

Their second trip north was no better. They found fossils – just not the kind they were looking for. And their time in Australia was running out.

Dick Tedford wrote in his journal that their trip was 'within 10 days of being a complete failure' before they were saved by an extraordinary bit of 'bone-digger's luck'.

Looking to escape the relentless harassment from sticky little bushflies, the team climbed up onto the windy crest of a sand dune to eat their lunch. The view from the top revealed a previously hidden salt pan lake. They had found Lake Palankarinna.

This site finally led to the diverse array of modern Australian mammal fossils they had been hoping for, to the great relief of the team. But even after they had finished and were ready to pack up their treasures to make the long trek back to the city, Paul Lawson was sure they had missed something. He went back out to an area they had already searched.

'A lucky blow from his geology hammer rolled out a tiny koala jaw,' recalled Tedford. 'This became the type specimen of the oldest known koala, *Perikoala palankarinnica*.'

You might think there is not much to tell about an animal from teeth or jaws. Teeth are the hardest and most mineralised part of a skeleton so, like shells, they are disproportionately common in the fossil record. In some species, teeth are fused to the jawbone, which survives along with them. Fortunately for scientists, teeth are also highly distinctive between species. Because they are vital for survival, evolutionary pressures for them to adapt and change are particularly strong.

Stirton's paper on *Perikoala palankarinnica* contains no photographs, but instead features crisp ink drawings of the teeth, which look like geological maps – a profile view of a mountain range on the horizon or an aerial shot of volcanic craters on a long-dead moon. Stirton lines them up alongside their modern counterparts, comparing the old with the new – the ancient possums with the modern ones, *Perikoala* next to *Phascolarctos*. The family resemblances are clear. These teeth reveal not only what is related (or not), but also how big it was and perhaps what it ate.

Perikoala palankarinnica weighed about as much as a domestic cat and was similar in size to the smallest modern koalas. Its jaw was short, suggesting it also shared their characteristically short snout or flattened face, rather than an elongated muzzle like a dog or kangaroo. Its teeth were simpler and flatter than modern koalas, indicating that it had a fairly general diet of softer leaves, without any tougher material. Since that first discovery, another species has been found in the area, the slightly smaller but more solidly built *Perikoala robustus*.

I am trying to imagine what this animal would have looked like, but struggle to conjure anything from these small fragments. How much like a modern koala was this creature? Was it just a smaller version, or did it look quite different – more like a possum, perhaps? My imagination produces only pale shadows of the animal I know from life, slight misshapen. The image elicits a weird sense of the uncanny valley, but offers me no insights.

I am struggling to give them character and depth, to make them as distinct from each other as modern species are. It requires more than different shades of brown or grey, spots or stripes. I need more detail if I am to attempt 'to consider a very ancient and long extinct group of mammals not as bits of broken bone but as flesh and blood beings', as the eminent palaeontologist George G. Simpson said in 1926.

These fragments cannot tell me if this animal had the short, strong limbs of a climber, with hands and feet for gripping, rather than those of a runner or walker on all fours. Did it have characteristic koala hands, with three fingers and two thumbs? Or their feet with two toes fused into a strange double-clawed digit, and one large stubby thumb that has lost its claw altogether? Teeth alone cannot reveal anything about the rest of the animal. It's like a stick figure of a koala with half the lines rubbed out.

I abandon my efforts at character and try world-building instead.

Twenty-four million years ago, a vast expanse of swampy forest spread across and around the shallow waters of Lake Palankarinna, supporting a diverse range of animals adapted to exploiting life in the trees and the water. Alongside the ancient koalas were an assortment of different possum species. Small carnivorous marsupials, not unlike modern antechinuses or shrews, hunted for insects in the branches. The waters were home to toothed platypus, otherwise similar in size and shape to the modern platypus. Zygomaturines, delicately built animals about the size of a miniature goat, browsed on the vegetation. Zygomaturine fossils are often associated with wet, swampy forests, although whether they were semi-aquatic remains contentious. Other species here were more commonly associated with life on land – a couple of dog-sized, wombat-like herbivores, a miniature kangaroo and a relative of the bandicoots. Fossils of predators are less common, but the recent rare find of a large forest hawk (*Archaehierax sylvestris*) at nearby Lake Pinpa suggests that the most significant threat to ancient koalas came from the sky.

I remember camping in the Otway forests, inland from the Great Ocean Road in Victoria. The fine-leafed southern beeches that once dominated the Oligocene rainforests are interspersed with the more modern eucalypts like messmates and manna gums. Tall shady tree ferns rise from a forest floor covered in moss and brackens. The landscape here is still filled with swamps and creeks and shallow lakes that dry periodically to reveal megafaunal trackways across their surfaces. There are no traces of ancient koalas here, but their modern descendants are abundant. It's as close a model as I can imagine for the swampy forests that once surrounded the now hot, dry lakes of the South Australian inland.

Mada means 'new' in the language of the Dieri people, the traditional First Nation owners of the land to the east of the Lake Eyre basin. After centuries of overwriting the landscape and animals with European names, recognition of, and collaboration with, Indigenous knowledge-holders and experts is gradually increasing in the scientific literature, even if it is in the mangle of Greek/Latin/English used for scientific nomenclature.

Madakoala, a larger species than *Perikoala*, was described from a collection of teeth, along with parts of the jaw and a crushed section of skull, first collected by Tedford at Lake Pinpa on a return visit to Australia in 1971. When I finally track down a published description of this species, I notice photos of the *Madakoala* and *Perikoala* teeth are lined up in pairs, and it takes me a while to realise they are intended to be viewed in 3D. I find an old pair of reading glasses, cut them in half and reposition them at the right distance apart to form a makeshift stereoscope.

The teeth pop out on the page in crisp clear focus. The 3D view reveals ridges that are sharper and steeper on *Madakoala* than *Perikoala*. It's something you could probably feel more than see. I find myself running my tongue over my own teeth for comparison. They feel like they have sharp ridges too, but human teeth are gentle hills compared to these rugged mountain ranges. These ones must have been like blades.

There's only so much that pictures, papers and photos can tell you. Sometimes you just need to see the real thing.

Palaeontology labs always seem to have something new and exciting underway: an interesting find, a new technique or a major discovery. Cupboards, shelves, drawers and benchtops are laden with fossils. Strange bits of bone or rock poke out from blankets, sit in trays, under microscopes or in buckets. Fossils often remain wrapped in boxes for years or decades, waiting for someone with the specific expertise to pick them up, turn them over and notice something unexpected or unusual. It takes long hours of patient preparation, measurement and analysis to reveal the significance of a find. The interesting thing about fossils is not so much the objects themselves, as the stories told about them.

Reconstructing the prehistoric animals of the past as they might have lived, breathed and behaved is an almost impossible task. We can reconstruct fragments of their physiology from the fragments of bone we find. We can make a guess from living relatives. And we can find clues in rare footprints, chemical signatures in the rock, soft tissue traces, eggs, nests, claw marks, scats or other remnants.

For some species, the fossil record is rich indeed. We know exactly what mammoths ate from the food frozen in their mouths and guts. We know what ate elasmosaurs from imprints of shark's teeth embedded in bone. We know which nerve cells sparked in the brains of ancient fish and the colour of pterosaur feathers. And sometimes we know which animals gave birth, or where they laid their eggs, or how long their young remained in a pouch.

For ancient koalas, though, we have very little – rarely more than a tooth or jaw fragment – to reconstruct an entire species. And so we just assume, not unreasonably, that a fossil koala species would have looked more or less like a modern koala albeit with variations in size. Prehistoric koalas are nearly always illustrated as looking much like modern koalas, one a little greyer, another more a rusty brown.

It's fair to say that members of the same family do bear a strong family resemblance. Canidae are dog-like, Felidae are cat-like and most Caviidae look like guinea pigs. But there are significant differences too. The tooth of an unknown prehistoric canid might turn out to be like a fox, a bush dog, a racoon dog or a wolf, or some quite different variation on the canid form. The hominid family includes bipedal running humans, arboreal orangutans, gigantic ground-dwelling gorillas and climbing quadrupedal chimps. If we had more living species of koalas, we might be more imaginative about them too.

What would we find if we discovered an entire prehistoric koala skeleton, or even something more than a few teeth? A different kind of leg, arm, hand, foot, spine or rib might suggest an animal that walked on the ground, dug holes or leapt and swung – maybe even gliding between trees. Were there prehistoric koalas that were more like wombats, more like possums, or more like something else entirely?

Did all ancient koalas sit in trees, using their hands for manipulation as well as for climbing? Or did they have a more quadrupedal stance? Were their ears small or large, round or tufted? Was their fur striped or spotted in shades of brown, yellow or red? Were they always tail-less like wombats or did they have tails like thylacoleos or diprotodons? And if they did have a tail, when did modern koalas lose theirs?

A skull might tell us if other koala species had the same small eyes or large noses. Did early koalas have the slightly flattened face of modern koalas, or the longer muzzle of many possums? A big part of the appeal of koalas lies in the shape of their face. Deep in our amygdala, humans have special facial-recognition neurons that respond to faces, particularly those of children, by releasing oxytocin. Did ancient koalas also have a face that a human could love?

The chances of finding a complete skeleton of a koala are remote. But if someone did, it would rewrite what we know about koala evolution. Rather than just comparing animals on the basis of their teeth (as a great deal of palaeontology does), we would have a full template for all the bones in the species, allowing other fragments already collected to be identified and allocated to their species or genus. We'd know more about how different body parts scale in size with changes in tooth size and be able to estimate body size more accurately.

Perhaps somewhere out there, on the edge of a dry lakebed, at the bottom of a cave or in the sediment of a flood plain, the conditions were just right for that rare preservation of a complete skeleton to lie, undisturbed, for long enough to be transformed into stone. And long enough for it to be uncovered by wind, rain, pick or shovel, sent to a palaeontology lab and patiently prepared, examined, written up and published.

The odds are not great, but you never know. We'll just have to hope we get lucky.

On our way home from the far north, we stop at Woomera, where I'm giving a talk at the local school. Woomera is a small town built in the 1950s in the middle of nowhere to service the space launch site. It also happens to be where my parents first met, and where I spent the first year of my life.

The school feels big enough to house a hundred students, but I have already been warned that there are only twelve enrolled. On the day, my class is reduced to one, as the others are all home with a cold. The town looks like a 1960s TV set – impeccably maintained but almost empty. Next to the café and tiny supermarket, I notice a small exhibition about the local area. In one cabinet there is a rock, impressed with characteristic leaves and fruit of trees. They are distinct impressions of gumnuts and gum leaves from the eucalypts that now dominate Australia. I suddenly remember that in 1962 Tedford had also found a leaf impression that looked much like eucalypt leaves. The first record of eucalypt pollen appears in the fossil record about 24 million years ago – at much the same time as the first koalas do.

4

From the Gulf to the Sea

Riversleigh, in the far north of Australia, always strikes me as a slightly magical place. It's partly because the name reminds me of Rivendell – J.R.R. Tolkien's otherworldly Elvish sanctuary. Like Rivendell, Riversleigh is a remote and hidden place – its secrets are concealed within rune-marked rocks in a long-forgotten language yet to be fully deciphered.

The Boodjamulla (Lawn Hill) National Park, which encompasses the Riversleigh site, is a spectacular landscape. Gaunt ochre escarpments and plunging gorges fracture a green-grey sea of dry eucalypt woodlands that stretches across the vast inland Gulf Country, on Queensland's border with the Northern Territory. But hidden beneath this surface lies another world – a lush, lost world that is 15–25 million years old, of rainforests and lakes rich in strange creatures both similar to, and quite unlike, our own.

The Riversleigh World Heritage site is notable for its rich diversity of mammal fossils, revealing much about the evolution of Australia's distinctive fauna. Giant flightless mihirungs and emuaries lived alongside cleaver-headed crocodiles, carnivorous kangaroos, toothed platypus, and marsupial moles, along with a wide range of ancestral kangaroos, bats, possums, thylacines, marsupial lions, wombats and, of course, an abundance of koalas.

'It's the epicentre for koala diversity in Australia,' explains Dr Gilbert Price, a vertebrate palaeoecologist and geochronologist from the University of Queensland. 'There are so many fossil species of koalas found in Queensland.'

Gilbert's research focuses on the ecological impacts of climate change over the last two and a half million years. Understanding the factors that have sent closely related species extinct in the recent past is surely essential for better protecting threatened species today. The distribution and extinction patterns of fossil wombats, Gilbert explains, demonstrate that rainfall, or water supply, has been crucial for their survival. It makes me wonder what patterns we'd see for extinct koalas.

Despite their former abundance, koalas are not so common in Queensland any more. I don't remember seeing any koalas when I lived near Cairns, at the most northerly end of their modern range. They are occasionally reported as far north as Cooktown, but more often they're on the dry inland side of the ranges, where the eucalypts outcompete the rainforest. The koala range has contracted and their numbers have reduced greatly.

'I grew up in bushland west of Brisbane,' Gilbert tells me. 'I've only seen two koalas in the wild up here in my whole life. They are just really rare. And the fossils are rare too. It wasn't until I found a fossil koala tooth that I really became interested in them.'

At least three or four different genera of koalas have been identified from Riversleigh. Species of *Litokoala* were about the size of a brush-tailed possum (or domestic cat) with large eyes, which suggests they were nocturnal. They also had incredibly complex teeth: 'a multitude of cusps and accessory shearing blades'. These teeth more closely resemble those of the possum-like cuscus species of the tropical rainforests than modern koalas – indicating that *Litokoala*s were specialised for eating nutrient-rich plants of Miocene rainforests, including seeds and fruits as well as leaves.

The oldest of these koalas, *Litokoala kutjamarpensis*, seems to have been quite widespread, as it has been found across several different locations in Riversleigh and also in South Australia. Two other similar-sized *Litokoala* species have also been found in Riversleigh, and they shared their forest with smaller and larger cousins. *Priscakoala* was similar in size to the smaller modern koalas (about 5 kilograms), but with relatively simple smooth molars. *Litokoala* and *Priscakoala* were uncommon compared to a smaller cousin, *Nimiokoala greystanesi*, which had complex molars adapted for eating tougher leaves. This small possum-sized koala accounts for one-third of all the fossil koalas found at Riversleigh.

Several of these species seem to have lived alongside each other in the same rainforests. For species to evolve and speciate into different animals, they must be separated by time, space or behaviour. If these koalas coexisted, how did they segregate from each other in the rainforest trees? How did they avoid interbreeding and competing with each other?

Perhaps the larger *Priscakoala* munched on the pick of the softest, easiest rainforest leaves during the day, curling up to sleep and conserving energy and warmth at night. *Nimiokoala* might have used its small size and greater agility to access foliage out of reach of *Priscakoala*, while its more complex shearing teeth allowed it to eat tougher tree species than the larger animal could. *Litokoala*, on the other hand, presumably kept out of harm's way during the day, exploiting the colder night-time forests to avoid its bigger and more abundant cousins.

There was one more koala from Riversleigh that we know very little about. *Stelakoala riversleighensis* is known only from a single tooth, which is larger than those of the other Riversleigh koalas and has a distinctive pattern of buttressing and ribbing more like those of modern animals. It's still at the small end of the size range for modern koalas (perhaps about 6 kilograms) but the similarity of its tooth to the modern species has led scientists to speculate that it might be the missing link between *Litokoala* and the modern species. There are no certainties, though. The koala fossil record, like that of humans, is patchy and broken. We like to join the dots between species across millions of years. Primates are notoriously impatient with puzzles we cannot solve.

I ask Gilbert about what the teeth tell us about the changing diet of koalas.

'Nearly all of the other fossil species have really simple teeth – nothing like as complex as the modern koala. They were probably eating softer leaves of rainforest trees. The modern koalas clearly had a really different diet of much tougher leaves.'

Like the modern koalas, *Stelakoala* was eating tougher, more fibrous food than its counterparts. *Stelakoala* lived at the 'climactic optimum' of the middle Miocene, 12–16 million years ago, when the Riversleigh rainforests were at their peak. Over the next two million years, the temperatures plunged by as much as 7 degrees Celsius, the rainforest species declined and contracted towards the coast, and the eucalypts rose to dominance. Australia had begun a progressive shift into the increasingly arid and seasonal climate we have today. This changing climate would dramatically transform the nature and extent of the forests covering the eastern and southern regions of the continent. Koalas would have to adapt too if they were to survive in this new forest world.

We can see the impact of this changing climate a thousand kilometres to the south-east of Riversleigh, in the opposite corner of Queensland, in the Chinchilla Sand fossil deposits of the Darling Downs. In the middle of the Pliocene, 3.6 million years ago, the climate here was 3 degrees warmer than today, the woodlands were contracting and the first significant grasslands appeared, supporting animals that grazed as modern kangaroos do today. The expanding grasslands would have held little interest to tree-dwelling koalas, but any change in the forests would have had a major impact. There were still rainforest-eating koalas around. The later *Invictokoala*, whose fossils were found 500 kilometres north in the Mount Etna Caves near Rockhampton in Queensland, was about the size of a modern koala, but its simple teeth suggest that it predominantly ate soft rainforest leaves. But other koalas were showing signs of adapting to tougher food. These were the *Phascolarctos* koalas – the immediate family of the modern koala, *Phascolarctos cinereus*, that we know today.

Ever since the Queensland Museum curator Charles de Vis first described his giant fanged koala in the late 1880s, these diprotodont fossil fragments from south-eastern Queensland had remained enigmatic. For example, de Vis originally described one of the fragments, *Koobor*, as a possum, but it has variously been moved to the extinct wombat-like Ilariids, to its own family alongside the wombats and koalas, or within the koalas as the closest relative of *Priscakoala*. Few of the original specimens were even labelled with a precise location.

'De Vis had a mountain of material to classify,' Gilbert Price explains. As one of a handful of scientific experts in a country filled

with new and unfamiliar species, de Vis had his work cut out bringing
any kind of order to the museum's collection.

'It was a bit like throwing a dart at a board with a blindfold on,'
Gilbert continues. 'He had a 50 per cent hit rate.'

It wasn't until 1968 that anyone looked closely at de Vis's giant
fanged 'koala'. By that time, many more fossils had emerged from
south-eastern Queensland, in an 800-kilometre line stretching along
the inland side of the Great Dividing Range from the New South Wales
border to Rockhampton. Following in de Vis's footsteps as curator
and then museum director, Dr Alan Bartholomai noticed that these
fossils were heavily mineralised and stained by iron oxides, characteristic of fossils from the Chinchilla Sand area, which date from 2 million
to 3.6 million years ago.

The surviving patches of native vegetation across this region today
are forests and open woodlands of acacias and eucalypts interspersed
with grasslands. In the past, the vegetation was more complex: a
mosaic of wetlands, grasslands and tropical forests supporting a rich
diversity of species, of which eucalypts only formed a minor part. The
seasonal wetlands were home to turtles and crocodiles and perhaps
wallowing zygomaturines, with an abundance of birds populating the
shorelines. The expanding grasslands suited emus, grazing kangaroos
and hunting falcons. Reptiles like goannas, skinks, snakes and even
komodo dragons warmed themselves in sunlit patches, while the forests sheltered quolls and thylacines, bandicoots and wombats, smaller
diprotodontoids like palorchestes, short-faced kangaroos, wallabies
and a diversity of rodents and bats. High up in the trees lived cuscuses
and a variety of koalas.

Alan Bartholomai decided that the mysterious shinbone de Vis
had described was indeed koala-like but, at twelve times the size of

a modern koala (and with key differences in structure), it was more likely to belong to a palorchestes or another member of the diprotodontoid family. The skull fragment, though, was definitely like a modern koala – but about 50 per cent bigger. Bartholomai thought this jaw fragment seemed very similar to another, more complete, oversized fossil koala jaw from Cement Mills, near Gore in southeastern Queensland, which he named *Phascolarctos stirtoni*, after the pioneering Professor Stirton. Similar fragments had also been found in Mammoth Cave in Western Australia, suggesting a very wide distribution.

Phascolarctos stirtoni was certainly hefty for a koala, but not everyone was convinced that it was a separate species to the modern *Phascolarctos cinereus*. Many animals seem to change their size over time, but does size alone make them a different species? Is the megafaunal kangaroo *Macropus titan* a different species to the modern eastern grey kangaroo, *Macropus giganteus*, or just a larger version of the same species? It's one thing to define species at a single point in time, but it's another to decide what constitutes a separate species from fragments collected at different points in time, perhaps even during the process of speciation. While Bartholomai also noted significant variations between the teeth of the fossil *Phascolarctos stirtoni* and modern *Phascolarctos cinereus*, these differences are not as pronounced as those between other fossil species. Is it possible that the koala is what some scientists call a 'time dwarf'?

It's a question Gilbert Price has asked too and tested more thoroughly.

'The teeth are around 30 per cent bigger than the modern koala and the ridges on the sides of the teeth are quite different,' he says. 'We have a lot of these fossils to compare now, and it looks like these two species coexisted in the same place and time – there's no evidence that the larger one evolved first and the smaller one came later.'

'So how do you work out how big the animal was just from its tooth?' I ask Gilbert.

'There's a formula for calculating body weight, based on several measures of tooth size across all the marsupial groups,' Gilbert replies. 'It's only a general estimate, but I've crunched the numbers and the weight comes out at about 23 kilograms.'

Two to three times bigger than a modern koala. More bull terrier than fox terrier. Not quite a giant, but a substantial animal all the same.

Maybe there is some physiological limit to how big climbing animals can get – like flight in birds. Maybe the trees just aren't strong enough to support a giant koala?

Tree climbing is certainly easier for smaller animals, but I'm not sure size is a major constraint. Larger apes (like chimpanzees and gorillas) do tend to spend more time on the ground than smaller ones, but plenty of larger animals easily climb trees, like leopards, cheetahs and jaguars. Most bears are adept climbers, from the smallest and most agile sun bears of 60–80 kilograms to the black bears and grizzly bears that weigh up to 200 kilograms.

And there have been larger tree-climbers in the past too. The 100-kilogram carnivorous *Thylacoleo carnifex* was clearly capable of climbing, not only trees but also cave dens where it raised its young, leaving an abundance of scratch marks on the steep walls leading out. The 68-kilogram diprotodon *Nimbadon lavarackorum* had flexible, rotating joints and short powerful limbs with gripping clawed feet, typical of tree-climbers. And there was even a giant tree kangaroo, *Congruus kitcheneri*, weighing up to 60 kilograms and standing 1 metre tall, in the forests that once covered the now famously tree-less Nullarbor Plain. Australian eucalypts include some of the tallest and hardest trees in the world. I'm certain they could bear the weight of a koala of more generous proportions.

Surely, somewhere among all the megafaunal creatures that existed in the past, there must have been a giant koala too?

5

A Giant at the Foot of the World

The road down the centre of Yorke Peninsula stretches south in a long, unwavering line towards a distant vanishing point. Sweeping plains of golden wheat to the east and west overlay bones of white limestone, interspersed here and there with silver salt pans that shimmer, mirage-like, in summer heat. On either side, the sea and the sky encircle a blue horizon. The only trees in this landscape cluster along the road verges, thick and scrubby, rarely more than 6 metres tall. I love this place, with its open vistas – it is the landscape of my childhood – but it is hardly the place I'd expect to find a koala.

I was surprised to read that koalas once lived here. When I first read the name *Phascolarctos yorkensis*, I assumed this fossil koala must have been found in Queensland – in the lush rainforests of Cape York, at the most northerly extreme of Australia. In fact, it was discovered much closer to home on Yorke Peninsula, not far from the popular holiday fishing destination where my grandparents lived for many years, just a few hours from Adelaide. I had no idea that there were any caves in the area, so I ask Gavin Prideaux, one of the local palaeontologists at Flinders University, how I could find it. As it happens, he knows the site very well – his wife's family once owned the farm where the cave is located.

'Corra-Lynn cave is on private land,' Gavin tells me. 'You can only go with a caving club. There's one at the university that does regular trips. You won't be able to go where the fossils are found, but it will give you a good feel for what the location is like.'

It seems like a good idea. I'm not planning a full-on caving expedition. I just want to see what the landscape is like – to try to imagine what it once would have been when the wide open wheatfields were covered in forests. Perhaps I can tag along on a caving trip, maybe peek inside the cave entrance and have a wander around while the serious cavers do their thing.

I send some preliminary emails, on the off-chance that there might be a trip at some stage in the next few months, and am swiftly invited to join a caving expedition the next weekend. A colleague generously provides a crash-course in caving basics over coffee, along with an assortment of helmets, torches, kneepads and first-aid kits. I eye them warily. It seems I am getting in deeper than I thought.

'You can turn back at any time,' she assures me. 'Anything you are not comfortable with, just say so and they'll take you out.'

Another caving friend advises me solemnly, 'Always look backwards, so you can know how to get out. And always carry two sets of spare batteries.'

I find this advice rather unsettling.

The leader of the expedition is Graham Pilkington, a veteran caver in his seventies. On the drive there, he tells me cheerily about all the groups of schoolchildren he has taken through the caves recently and how the biggest problem was always the teacher. I'm not sure if that has to do with courage or common sense, but I can see that the prospect of being able to bail out with any semblance of dignity is becoming increasingly unlikely.

The cave entrance is concealed in the middle of seemingly flat farmland. A cleft in one side of a gently sloping hillock opens into a dark chasm that leads to a small steel door, heavily bolted and locked. We walk down dusty steps into a narrowing corridor, our heads soon bumping on the hard roof. If I was expecting a vast cavern to wander through – a welcoming antechamber before the labyrinth – I am quickly disabused. The cave peters out at an unpromising apex in which only a few dark corners and shadows offer slim hope of access. As we wait for our small party to gather, I notice an opening to the side – hobbit-sized at best, but not too tiny. But Graham turns the other way, and promptly disappears down a small hole in the ground that I am sure even a wombat would view with scepticism. As his feet vanish, I realise with some horror that I am next in line.

'You okay to go next?' asks Sarah, an experienced caver behind me. Best not to think too much about it, I decide. I take a deep breath and dive in.

The tunnels are smaller and longer than I expected. Barely big enough for crawling, in many spots there is only space to wriggle, arms outstretched, with boots scrabbling for traction. From time to time I have to turn my head to fit my helmet through. Claustrophobia is not an option. I remind myself, over and over, that I am the smallest person in the group, that hundreds of people have worn these tunnels smooth, and that if you can get in, you can get out. There is really no logical reason why you would get stuck. It becomes a mantra to stop me thinking about other possibilities.

I also remind myself that primates are, underneath it all, still climbers. Like koalas, we are adapted for clinging, clambering and climbing: opposable 'thumbs', friction-ridged fingerprints, long limbs, a short body and rotating shoulder joints. The adaptations for tree-climbing are surprisingly similar to those for rock-climbing. There's not as much difference as you might think between a tree kangaroo and a rock wallaby. I focus on one step at a time, one tunnel at a time. My sedentary, desk-bound body is not accustomed to such exercise, but there is a satisfying rhythm to the effort which distracts me from the endless dark.

Periodically we emerge into narrow rifts in the rock that open into darkness above and below. I realise that the smaller tunnels have been dug out by hand to join up these different caves – which explains why the tunnels are only just big enough to allow human access. We sit in a rare spacious corner, the beams of our head torches creating a crosshatch of light in the darkness, as we catch our breath and wait for the rest of the group.

I hadn't expected caves to be shaped like this.

'These are dry caves,' explains Sarah, the geologist on the trip. She shows me a map of the cave structure: an irregular multi-layered houndstooth pattern labelled with idiosyncratic names like 'The Walrus' and 'Beard Squeeze'. 'The caves were formed by the water table rising and dissolving the softer rocks that had filled the fractures between older harder rocks.'

She adds, 'There are probably other cave systems underground that we don't know about. Farmers tend to fill in surface holes so that stock can't fall in.'

I remember playing hide-and-seek in the limestone potholes in the paddocks where I grew up. The same kinds of holes could easily have opened into a labyrinth like this one, where countless generations of animals have fallen to their deaths, their bones littering the cave floor. Only a few caves, with openings to the surface, collect fossils like this.

'How far is it to Dreamland?' someone asks. It's where all the fossils have been found.

'Two hours in and two hours out,' Graham replies. 'Pretty tight in places.'

'Tighter than Wombat Run?' someone else asks.

'Worse than Bandicoot Bypass,' comes a gloomy voice from the darkness.

Graham makes a dismissive noise. This is a man who has opened up large sections of this cave system by digging out the ends of tunnels where he thinks they might connect to other systems, carrying out the dirt in his pockets and a lunchbox because the tunnels are too small to accommodate any backfill. He's squeezed himself into crevices so tight you have to breath out in order to get in and then scrape out enough dirt to give yourself room to take the next breath. Not that long ago, he discovered that he'd broken some ribs in the process.

'You have to get through the Letterbox first,' says Graham, 'then through Alberta, which is fairly tight and long, so you get tired by the end.' He points to a section of the map illuminated by his headlamp. It's labelled 'The ##*!'

Graham continues, 'And then it opens up to the Portal. That's the slow spot.'

The Portal is a 7-metre vertical climb through a squeeze between the different levels of the caves. It must be done slowly, one person at a time. It's the only way to reach the fossil caves – The Bench, Koala Patch and The Graveyard. The latter is accessible through a tunnel so narrow that one of the last cavers to enter it had to strip off his overalls, almost naked, to get through.

In the 1980s, Graham and others found bones in the pockets of red sand here: frog, snake, bird and marsupial rat bones, as well as koala teeth and jaws – all perfectly normal and unexceptional. But in 1985, three of them finished off a busy weekend exploring with one final visit to a previously unsurveyed cave. In the cemented red sands that had collapsed within the cave, they found an older koala jaw – one at least double the size of modern koala jaws. This creature would have dwarfed the biggest male koala around today. It was originally named *Cundokoala yorkensis*, or 'thunder' koala from the local Narungga language, but was later reclassified as *Phascolarctos yorkensis*.

'That must have been pretty exciting,' I say to Graham.

He shrugs.

'It's good when you find something new,' he says, 'but fossils are a bit a nuisance, really. They stop you from digging new tunnels.'

When we emerged from the cave, the world was suddenly full of light, sound and movement. A stiff breeze blasted across the flat landscape, strangely disorienting after the dark, muffled stillness below. I emerged with a whole new respect for the words I so often read in palaeontology papers when they mention in passing that 'cavers found this fossil ...'

'Did you find the fossils?' a friend later asks me excitedly.

'No, that wasn't the plan,' I explain, but she still looks puzzled and crestfallen. 'I just wanted to get a better understanding about where the koala fossils were from and how everything fitted together.'

'Reckon you could have done that without spending five hours crawling around in a hole,' she replies. Maybe. But I get a feeling that my questions about koalas are not going to be easy to answer. Some things have to be done the hard way.

The limestone that underlies Yorke Peninsula periodically breaks through the shallow soils in gleaming white sheets. It comes to the surface in misshapen lumps, under the plough, flung in piles that line the edges of wheatfields like bleached bones in the sun, more rubble than dry stone walls. This labour of past generations reminds me of just how modified this landscape is, how different it must have been before Europeans cleared the trees, dug up the rocks and drained the swamps.

It's only on the southern tip of the peninsula that I appreciate how vast this difference is. A row of brackish swampy lakes isolates the toe at the foot-shaped tip of the peninsula. The remains of old drainage works and gypsum-mining activities mark the efforts of former residents to develop the area. But the harsh coastal conditions, poor soil and extensive lakes thwarted their efforts, and in the 1970s the area was declared a national park.

Looking at the swampy lakes, I remember a PhD student who took one of my classes in academic writing. He had described the fragility of the freshwater aquifers under these peninsulas and the incursions of saltwater from the ocean. The ground water beneath these lakes is increasingly salty, at risk from bores that disrupt the delicate tensions holding the water in place, like the surface membrane of a droplet of water balancing on a tabletop.

It had never occurred to me before how much this underground water shaped the landscape and how important it has been. Once, when the watertable was higher, it dissolved the soft fingers of rock that had infiltrated cracks in the ancient Cambrian limestone, creating the Corra-Lynn cave formation. Earlier still, it had filled the low-lying areas of the peninsula with shallow freshwater lakes, which had supported the forests that once covered the southern areas of Australia and the prehistoric koalas that inhabited it.

These long-vanished swampy forests keep recurring in the koala story. Wherever there are koala fossils – across different ages and locations – there were swamps and forests. It would be easy to assume that this is purely because swamps are good for making fossils. If koalas were found in drier forests, they would not have left fossils for us to find.

But I can't help noticing that the favourite food trees of koalas today – like manna gums, swamp gums and river red gums – are all 'riparian' species. They are trees that grow in damp areas and along the banks of rivers. As a result, koalas now are almost entirely distributed along watercourses, floodplains, inland rivers and wet forests, just as fossils of their predecessors were. It's an interesting coincidence.

'Was *yorkensis* really a giant?' I ask Gavin on my return.

It's always tempting to want to find the biggest animal, a giant. I should know – it's what I'm doing now. But Gavin works predominantly in megafauna, and he's particularly sceptical about any unsupported claims of giants.

'Oh – this one's about three times bigger and much more robust. It's a very different beast to *Phascolarctos*,' says Gavin.

He shows me the jaw of a modern koala, and then the fossilised jaw of *Phascolarctos yorkensis*. It is huge – a thunder koala indeed. I am struck by the teeth. They are shiny and white, as if they are still enamel within pale bone, with a delicate shadowing of cyan blue highlighting their scalloped selenodont curves. They are unexpectedly pretty – immaculate rows of overlapping crescents, like cockle shells all in a row, or some kind of delicate lacework. I turn them on their side and realise that they remind me of the Sydney Opera House.

When I get home, I take down the koala skull on my shelf, which my stepfather gave me. He can't remember where it came from, but it was probably given to him when he volunteered at the Adelaide Zoo. I look more closely at the sharp, raised ridges of the teeth. The patterns are there too, but not as pretty. Dark residue stains the clefts like incipient cavities. They look like they need a good clean.

The skull fits snugly in the palm of my hand – smaller than I expect. The fur and ears of koalas are misleading. I'm reminded of the way my cat sometimes rests her head in my hand, and I'm surprised that something so small can be so full of life and personality.

The modern koala is only about the size of a two-year-old child – a comfortable weight and size for someone to carry. *Phascolarctos yorkensis* might have been closer to 30 kilograms, although it's hard to be sure from these fragments. More the size of a nine- or ten-year-old child. Maybe more like a baboon.

I realise that I am thinking about koalas as being 'like' again. Like a child, like a baboon. Like a child makes me think they are helpless and endearing. Like a baboon makes me imagine them as aggressive and domineering.

It's hard to guess the nature of different animals from how well armed they are, or what they eat. South American tapirs are largely amiable creatures, while smaller Malayan tapirs have a reputation among zookeepers for being highly aggressive and bad-tempered. Who knows whether the giant koala shared the generally placid nature of its modern counterpart or was a different kind of beast altogether?

Following the journey from small possum-like nocturnal ancestors in the soft-leaved, nutrient-rich forests of 24 million years ago to the giants of the south has helped me stop thinking about koalas as singular and unique creatures in a sea of unchanging eucalypts. I can see now how, as a group, they have been shaped through time by changes in climate and vegetation, by competition with others and by the challenges they have faced. How well this ancestry equips them to face the unprecedented threats of the future is, however, another matter entirely.

The taxonomy of koalas – their family history – is only half the story, though. I want to find out how to build a koala, from the bones up.

III

LIFE IN THE FOREST

⌒ At the top of the tallest tree, the koala is enveloped in a curtain of greenery, surrounded by the quiet hum of insects and the occasional discordant shriek of parrots. It has no reason to look beyond its immediate surroundings. The canopy stretches unchanging from horizon to horizon in a mosaic of olive-grey-blue-green. This forested realm seems endless, sweeping down the foothills, carpeting the plains where the darkened tracks of rivers drain vein-like into valleys, before meandering across the lowlands into the expanse of open water beyond.

Above the forest rise granite mountains: the igneous backbone of the eastern seaboard from north to south. The peaks frost white and snow gums entwine over open grass and heath lands. Shallow sphagnum bogs rest between submerged boulders, tucked beneath mossy blankets under frigid skies.

The koala rarely ventures up into the cold mountain forests. It lives on the lower slopes, where the tall forests reign. A kingdom of eucalypts: ash, box, bloodwood, messmate and manna, ironbark and peppermint. Monuments to photosynthetic power, the glory of chlorophyll: 200 tonnes of biomass rising in smooth, organic grey obelisks 20, 50, 100 metres high, conjured miraculously from carbon dioxide and water. The invisible and the transparent made mountainous by sheer solar radiance.

These forests descend from eucalypt ancestors whose pollen has dusted the Australian landscape for 25 million years. They began their indomitable ascendance as Australia drifted north. In the drier, colder conditions, the hard, waxy sclerophyll leaves gained an advantage. The eucalypt forests spread and contracted, radiated and speciated, adapted and diversified. Restricted to refugia by the Last Glacial Maximum, the forest that currently spreads across the montane zone – this particular, peculiar, carefully structured aggregation of species – has prevailed for 20,000 years, regulated by an increasingly powerful selective force. Fire has crowned the eucalypts ascendant in this dry, impoverished land.

They are citadels of biodiversity, these trees: each branch, each hollow, each leaf, each crevice, each crack, home to legions of species, sheltering, sleeping, feeding, breeding, living and dying in a continual interconnected cycle.

The solitary koala is not alone in this tree, but one of many. Gliders and possums sleep in dark corners, safe from the glowering eyes of powerful owls, while tiny antechinuses and feathertails cluster in colonies in cramped, low-ranking accommodation. Twittering bats pack-jam into crooks and crannies. Families of parrots and pardalotes return, year in, year out, like annual holiday-makers. Large cockatoos dominate the prime real estate of spacious penthouse hollows – sulfur-crested, yellow-tailed, pink, grey and black. Chattering flocks of small birds swoop and mob the koala with indignant cries as it munches on, oblivious to their outrage.

The spreading canopy creates patches of sun and shade beneath, micro-habitats for myriad skinks that occupy their allotted place as the light and dark shrink and stretch. Tree ferns arch their foliage in an arboreal cathedral across the forest floor. Shade-loving plants cluster in dampness with prehistoric ferns, delicate epiphytes, mosses and lichens. Fragile webs of fungi feed snuffling bandicoots.

And in the canopy, insects race the koala, feeding ceaselessly on leafy factories of light. Chemicals in the leaves slow the onslaught. A complex bio-synthesis of compounds carefully tailored to invertebrates' Achilles heels – terpenes, alkaloids and phenols. But still the armies come: beetles, sawflies and caterpillars – leaf-mining, sap-sucking, bark-chewing, skeletonis-ing, gall-forming.

The koala reaches for a leaf, sniffing carefully before biting the shaft and tugging it free of the twiglet. The leaf is astringent and the sun is too hot. The koala retreats deeper into the shade and sleeps, waiting for the cool damp night and the promise of rain. ᴥ

6

Anatomy of a Climber

The koala down by the dam hasn't moved for a couple of days. That's not a good sign. The trees on the dam wall are introduced southern blue gums, not more than thirty years old. Koalas often appear in these trees but rarely stay long. I worry that it might be ill or injured. It's hard to know with an animal that sleeps so much. By the time I walk down to check on it, it's coming on dusk. The koala is wide-awake. There's nothing slow or slumbering about its movements now.

The koala trots casually along a horizontal branch 10 metres up in the canopy, with a distinct frisk in its step. As I watch, it stands up to grab the branch above and then slides swiftly sideways like someone hanging washing on a line. I lose sight of it behind the main trunk, before spotting its head re-emerging higher up. I can see it sizing up the gap between this tree and another. Without hesitation it leaps at least 2 metres onto the other tree then continues on its way.

I didn't realise they could jump so far, that they could be so agile. I suppose the chances of a sun-loving diurnal terrestrial primate seeing a mostly nocturnal koala during the few hours it is active are vanishingly small. We simply operate on different planes and in different time zones.

I admire the easy confidence with which koalas traverse the highest branches of our trees. It reminds me of a childhood spent climbing trees, swinging from ropes and building treehouses. Children have a supple fearlessness – that perfect ratio of muscle and bone to body mass where energy seems effortless; that elasticity and lightness, unencumbered by the heavy, leaden-footed solidity and anxiety that comes with maturity.

Climbing takes effort now. I struggle to lift myself into fruit trees to prune long branches or collect high fruit. I look hopefully at some of the low-branching gum trees with rough fibrous bark for easy traction. I want to remember what it's like to sit at the top of a tree – to see the world from a koala's perspective – but I know I'm not fit enough to try to climb them.

My daughter and I decide to visit an inner-city bouldering centre. The walls are lower and the mats beneath them are thick and bouncy. But it's harder than it looks, grasping the smooth hand grips, finding footholds at full stretch, holding your weight as the wall tips outwards. My feet seem too big to fit the footholds. My arms struggle to lift my weight and my legs are too long. My centre of balance is all wrong. I need short muscly limbs, gripping feet and a good set of claws. As we complete each increasingly difficult climb, we drop to the ground, puffing with satisfied exertion, taking turns to recover our breath. After half an hour, we no longer have the strength to grip onto the wall.

'It's good exercise, though,' my daughter says. 'It really uses all your muscles.'

I can feel every muscle in my back and abdomen. My shoulders feel stretched, pulled out of their habitual computer-generated hunch. My arms and legs surge with unexpected blood flow. It feels like the first time I've used my whole body in years – as if this was what it was built for.

As I walked up to check the letterbox, I realised I was not alone. A large koala ambled beneath the trees alongside me. If his size and more Romanesque profile did not identify him as male, the dark oily streak of scent glands down the centre of his chest was unmistakeable. He picked up his pace fractionally at my approach, his attentive ears the only other sign that he'd noticed me, passing with practised ease under the fence and through to the road verge, where he vaulted swiftly up a sloping gum before coming to rest just a metre or two from the ground. The koalas here rarely pause to inspect a source of danger when they are on the ground. Only once they have reached the security of a tree do they feel safe to stop and examine the world around them – even when they are still easily within reach. You can almost sense their relief – or is it confidence? – when their claws grasp the hardy gum trunks, that path of ascendant freedom.

To live in trees is to be arboreal – of the trees – but not all arboreal animals are climbers. Birds, butterflies and other insects have the advantage of flight, while climbers must tackle the challenges of gravity and grip with divergent solutions. They slide, slither, fling, swing, glide, scramble, cling and clamber, each with their own distinctive adaptations.

Climbing animals have less muscle mass than ground-dwelling animals. A koala has a third less muscle than a similarly sized Bennett's tree-kangaroo. All kangaroos are pretty muscly, but when tree kangaroos returned to an arboreal habitat, their muscle mass dropped back down to much the same as koalas.

This is not because climbing uses less energy than walking or running. In fact, it uses more. Climbing stairs takes much more effort than walking along a corridor. But animals that live on the ground tend to move around a lot more. They live in a place crowded with competitors and predators and they must be able to find food and escape – sometimes fast.

Animals that live in trees have already escaped. Not everything can climb and live on leaves. Animals that can sustain themselves in the specialised world of the trees find themselves in a smaller, safer space and with less reason to move around. They have a food supply all to themselves and are protected from ground-dwelling predators. They can afford to sit and watch and eat, rather than being constantly on guard, ready to flee. Climbers have less muscle not because climbing is easier, but because once you're up in the canopy you can relax.

I wander through the carpeted corridors of the local museum, looking for the skeleton of the koala. It's not easy to find in this gallery of calcified articulations. Stripped of their superficial differences, back to their bare bones, the common body plan of the mammals becomes more evident.

As usual, I'm forgetting to look up – still stuck in my two-dimensional terrestrial plane. The koala is on a branch overhead, as if looking out over the middle distance. A koala's skull doesn't sit on top of its spine like ours. Rather, it hangs like a large box-shaped ornament off the end of a long neck and spine, curved like a crescent moon. It takes effort to balance a head on a long neck, and the koala's neck vertebrae are enlarged to take the load.

I am struck by the koala's massive arm and shoulder bones. A robust collarbone and wide, flat shoulder blades brace their shoulders front and back, providing the attachment points for large muscles. The humerus, or upper arm bone, is the largest single bone in a koala's body, much thicker and slightly longer than the femur of the upper leg. It makes me think of wheelchair rugby players, relying on vast upper body strength to carry the weight of their entire bodies. Or infant marsupials at birth – little more than a mouth on arms, all head and powerful shoulders, climbing to their mother's teat and clinging on for dear life. Appearance can be deceiving. Overall, the muscle mass of hind and forelimbs is about equal, suggesting that koalas do as much pushing as pulling.

The koala's legs, tucked in close to its body, are longer than I expected for a climber. Even so, you can see how they keep themselves close to their centre of gravity, reducing 'top-heaviness' and utilising the flexibility of their joints. Heading upwards, they hug tree trunks in a pincer grip between their long arms, gripping with strong claws, and hop with their back feet together. It's a similar motion used by traditional tree-climbers to scale coconut trees with only the aid of an ankle belt. When koalas really want to move, they are capable of short bursts of surprising speed. They bound along the ground, using their powerful back legs to propel themselves forwards in that character-istically diprotodontid hopping motion perfected by their kangaroo cousins. If necessary, they can fling themselves straight up the smooth vertical face of trees too, using the same bounding movement. As I slow down a video to watch this action, I realise that they are resting their entire weight on their back legs as they stretch both arms up and out for the next stride. They make the impossible look effortless.

There is an old story that koalas don't have ribs. Maybe this was told by keepers to enthusiastic koala-huggers to discourage them from squeezing too tightly. Or because koalas (like most animals) should not be picked up around their ribs but supported from underneath. Koalas do have ribs, one pair fewer than the thirteen of most marsupials, that form a more delicate and flexible cone-shape than the robust, boxy twelve-paired ribcage of hominid apes. What koalas lack in rib strength, though, they gain in a massively over-engineered lower back, with lumbar vertebrae two or three times the size of the ones supporting the ribs. The spine curves inward towards the pelvis and back legs. This is a back built for sitting – for up to nineteen hours a day.

I've heard that koalas have a 'sitting plate' in their lower spine that allows them to 'lock' their position in trees. But I can't see this feature on their skeleton. The vertebrae diminish beyond the hips into a tiny residual tail – just like ours. We've both lost our tails, it seems. Like a fifth limb, a prehensile tail grips and clings to branches, in possums, cuscuses, spider monkeys, Brazilian porcupines and many rodents. I wish we'd kept a prehensile tail. It would be useful to have an extra hand for grasping and balancing. I think of the spider monkeys I used to look after in the zoo and the way they'd sometimes steal food with their tails, or sit with arms, legs and tails wrapped around each other for extra warmth and affection. You'd think an extra hand would be helpful to a climber, but clearly not. Tails have been lost in many climbers – orangutans, the indri lemur, bears, macaques and humans. Perhaps a tail is simply a nuisance for animals that spend most of their time just sitting on it.

Both koalas and wombats are renowned for having unusually tough rear ends that make them nearly impervious to attacks from behind. The skin on a koala's rump is extremely tough and the fur particularly thick – it's regarded as almost bulletproof, and certainly a sturdy and comfortable cushion. The strength of a wombat's rear end is legendary. Kath Handasyde tells me that she often sees wombats running to their burrows and 'propping' in the entrance as if to block it with their rear.

I've read claims that wombats sometimes crouch down in the entrance of their burrows to entice would-be predators to push their heads in.

'I have two friends who've lost dogs to this behaviour – skulls crushed,' Kath says. 'I've occasionally tickled a wombat on the rump with a stick when it ran into the burrow and they definitely ram their rump upwards in response.'

Wom-butts of steel – as the biologists say – is really no joke.

This extraordinary defence mechanism is due to a 'sacral plate', variously described as being made of fused bones, cartilage or even toughened skin. But I can't find any clear anatomical descriptions for the koala, and even those for the wombat are unclear. My most authoritative source declares that the sacral plate of the wombat is 'almost as rigid as bone' but 'comprises thick, dense skin', as illustrated in an accompanying X-ray. I can only assume the same is true for koalas – that their 'bony' behind is, like the wombats, actually due to a dermal shield of super-toughened skin.

I ask David Stemmer, the curator of mammals at the South Australian Museum, whether he's ever noticed a bony plate or cartilage on a koala. Not many other people are likely to have dissected a koala before.

'I don't remember any unusual plates or structures,' he says, shaking his head. 'But the skin is incredibly tough and very firmly attached around the rump. It's not like other mammals.'

I wouldn't have thought skin would provide much protection, but dermal shields are not unusual in herbivores – the collagen-rich plating of rhinoceroses is an extreme example. I suspect that it is this dense, thick pad, and their very thick fur, that provides the koala its 'bulletproof' padding as they take advantage of the natural V-shapes in eucalypt branches, wedging themselves securely in the crevice between branch and trunk.

Koalas have perfected sitting in trees. They do not merely sit, as we do, in an unnatural posture on the edge of a chair, never sure where to put our long legs, how to curve our back, where to hold our shoulders. Koalas sit with ease and aplomb, and sleep with absolute confidence on their precarious spot. On hot days they cool themselves by sprawling belly-down across the highest branches, limbs dangling beneath them, or they laze back in nature's deckchairs with their feet propped up on a convenient perch. In winter wind and rain, they curl themselves into tight furred balls wedged between branches. They watch the world beneath, poised and regal on eucalypt thrones. They sit with gravity and firmness, not unsteadily, but leaning, lying, lolling, relaxing, resting, flopped, slumped or upright.

I straighten my back, somewhat painfully, from its usual slouching curve in my chair, victim of our imperfect evolution from quadruped to biped, and envy the semi-supine ease with which koalas effortlessly while away long summer days in the trees.

I watch as a koala descends backwards, its fur as grey as the trunk of the gum tree. It takes its time, ears turning, listening. Its powerful forearms are wide, taking the full weight of its body as it reaches the tree's broad base. Its short, curved back legs crouch beneath it like a backstop, ready to provide spring should any forward momentum be needed. At any point in its gait, legs outstretched or curled beneath, the koala's body stays hugged close to the tree, its weight centred and stable, never swinging. Its strong, sturdy claws splay black across the dense hardwood, distributing its weight evenly across its fingers and preventing it from rolling.

I'm impressed by how effectively those huge claws can grip such hard, smooth timber. They must be like razor blades at close quarters – particularly without the protection of thick fur.

'Koalas look so sweet, it's easy to forget that they can inflict some damage,' Kath tells me. She's had plenty of first-hand experience in forty years of studying koalas, both in the wild and in captivity.

'Those claws are so sharp you don't even notice when they get you. You think it's just a scratch until you notice the blood and realise how deep it is.'

Cute and cuddly, but with razor claws. More Wolverine than dropbear – but without the aggression.

'The cuts heal up super-fast too,' Kath reassures me, with the resilient good humour of a field biologist. 'Koalas don't come after you – you just have to pay attention and not get too close if they're upset or cranky. And who wouldn't be cranky after being brought down from their tree?'

I watch the koala's hand splay across the branch. It looks much the same as a human hand might. In fact, the similarity is striking – the same shortened palm with long fingers and opposable thumbs – despite the claws. It's a hand, not a paw. A climber's hand. But then I notice the fingers are split into a bipartite Vulcan salute – not four fingers and one thumb, like ours, but three fingers and two thumbs. I'm surprised I've never noticed this before.

Their feet are even stranger. In primate evolution, there have been endless debates around whether the other apes are four-handed, four-footed or have two hands and two feet, like humans, but the nomenclature seems more relaxed in marsupial circles. Koala feet technically have four clawed fingers and one stumpy, clawless thumb, but the first two fingers are very fine and partly fused. This 'syndactylous' conjoined finger with two claws is characteristic of almost all the Australian marsupials, except for the carnivorous dasyurids and the marsupial moles. What caused this feature remains a mystery, but most of the marsupials, like koalas, put their dual clawed toe to good use as a grooming tool.

Claws are not the only adaptation koalas have for grip. Their hands and feet are rough, like fine sandpaper. They are also, apparently, the only species to have fingerprints other than humans, chimpanzees and gorillas. Only the fingertips beneath their claws are covered by fingerprints, much less than the area on human or chimpanzee hands. I've heard that their fingerprints are so similar to human ones that if they were found on a crime scene, they would be included in the forensic examinations. And like human fingerprints, koala prints are unique to each individual.

Ridges on fingers and hands, feet and toes, and even tails provide grip for climbing, and many climbing animals have strangely grooved pads on the soles of their feet. The palms of koalas are covered in small round ridges, rather like the grip on a rubber gardening glove. But these are different to fingerprints. Fingerprints are tiny dermal ridges, knowns as 'dermatoglyphs' in the skin. These kinds of ridges are most pronounced on the fingertips, but they are also present in humans across the palm of the hand and on the feet.

It's not clear how common such fingerprints are in mammals. I find reports of similar fine dermal ridges in the common spotted cuscus and the Virginia opossum, and even a mustelid fisher 'cat' and the prehensile tails of some monkeys. The only common feature is their arboreal life.

It is easy to assume that fingerprints are just another adaptation for improving grip, but it's not certain that fingerprints increase friction. Friction is related to the amount of surface area connecting two surfaces. Smooth skin actually produces more friction than our ridged fingertips. But friction is an incredibly complicated formula, dependant on the texture, moisture and flexibility of both surfaces and the nature of the force exerted.

Fingerprints do, however, increase our sensitivity to touch. Sliding your fingers across any surface sets off vibrations in the skin, which allows you to sense minute changes in texture and shape. These vibrations are detected by mechanoreceptors called Pacinian corpuscles, which are embedded a couple of millimetres below the skin's surface. Vibrations can be detected with smooth skin, but the intensity of these skin vibrations is amplified about 100 times by the presence of fingerprints. I remember reading about violin makers who select timber by laying their hands on it and feeling the barely perceptible vibrations beneath their fingertips, which ultimately gives their instrument its personal resonance and timbre.

Maybe fingerprints have different purposes for different species, or maybe they have no particular function at all. Maybe they are just a development glitch in one type of foot pad. Perhaps there are so many various ways of achieving the same goal that there is no obvious pattern to them.

But I keep coming back to what we do with our hands – our ability to manipulate, grasp and feel. Our fine motor skills. And I think about the koalas, with their hands in constant contact with the tree they depend upon, reaching out with one hand to carefully select just the right leaf, sniff it and draw it to their mouth. And I can't help but wonder if there is something more specific in their world of touch and grip, finger to thumb, hand to tree – some sensation or communication that makes the delicate touch of a fingertip as important to koala evolution as it has been to our own.

7

The Eucalypt Empire

Bel sits calmly on her elevated throne, leaning back into the low forked tree stump that's been artistically decorated with fresh foliage. Her neatly uniformed attendant makes last-minute arrangements. We stand at the start of a well-worn path that winds its way towards her – corralled behind a gate to await our politely spaced five-minute audience with Australian royalty. Even this early on a mid-winter, mid-week morning, there is a small queue of visitors: mums with prams and toddlers, a few backpackers in bright anoraks, a studious-looking young man in glasses and an elderly couple. As we read the sign instructing us on appropriate koala protocol a sweet little long-nosed potoroo provides warm-up entertainment, hopping deftly between our feet in search of food. There is a murmur of quiet anticipation as the queue begins to move forwards.

Bel is fifteen years old and was raised from a joey at Cleland Wildlife Park. She's one of their star performers, placid and good-natured – a beautiful temperament for public engagements.

When it's my turn to approach, the keeper launches into her well-practised introduction on koalas, their ecology, diet and habits.

'You can pat her on the rump,' Cassie says. 'But please don't touch her head or ears. Would you like to have a photo taken with her?'

'Are they sensitive about their ears being touched?' I ask.

'Not particularly,' Cassie replies. 'Bel would be fine, but it's just a bit risky with the public. Sometimes people can startle the animals. We don't want them to be upset. If they don't like being here, they'll just get up and leave.'

Primates can be noisy and demonstrative creatures. Not all animals tolerate that kind of behaviour. I ask Cassie how they train the koalas for these public experiences.

'Very slowly,' she tells me. 'It takes a long time to get them used to feeding up close, being touched and being held. Not all koalas are suited to it. It's really important that this is always a positive experience for them.'

She hands Bel a fresh bunch of gum leaves.

'And is the food important?' I ask.

Cassie grins.

'Absolutely essential,' she says. 'This is *Eucalyptus platypus*. All of the koalas love it. We only use it as a special treat.'

Bel munches steadily on her eucalypt delicacy while I stroke the thick, soft fur on her back. Even with treats, she will only be here for thirty minutes at most.

'That's about enough activity for any koala on one day,' Cassie tells me.

My time is up. Cassie gives me some of the prized eucalypt treat to take with me before I head towards the exit sign, the next set of visitors moving up behind me.

As I drive home, my car is filled with the crisp, fresh scent of eucalyptus leaves. I've never seen gum leaves like these before – they seem artificial, as if they are made of plastic. They are not gnarly or hard or shades of grey like a lot of local gum trees. They are shiny bright green, with smooth, round light-green flower buds and neat trim little gumnuts. They almost look appetising. I wonder what makes this eucalyptus so irresistible for koalas. It's not like they'll just eat any old gum leaves. Koalas are notoriously fussy about their food.

Eucalyptus platypus is a Western Australian species, I discover, growing in a small region between Albany and Esperance and originally called *moort* or *maalok* by the Noongar people. It grows like mallee – low with a dense rounded crown on multiple trunks. Unlike most Australian mallees, though, it does not have the large underground root ball – known as a lignotuber – that makes mallee roots so popular for firewood, and so unpopular with farmers trying to clear their land. It seems strange that koalas would love a shrubby tree native to an area of Australia where they don't live.

I post a picture of the leaves online, making a joke about them being koala 'lollies' – Australian for candy.

'The leaves in the picture are *Eucalyptus utilis* (coastal moort), a close relative of *E. platypus*,' someone corrects me. 'Widely grown across southern Australia, often for shelterbelts.'

The coastal moort has a similar distribution to its close cousin but is slightly more widespread, and prone to being an invasive weed. Its leaves are narrower, more lance-like. It's not a distinction I would be able to make, but the correction has come from Dr Dean Nicolle, who owns the Currency Creek Arboretum. As the creator of the single largest collection of eucalypts grown in one place, Dean probably knows more about eucalypt species in the wild than anyone else in the world. Being able to accurately identify eucalypts in the wild, in all their various forms and ages, is his speciality. If I want to understand eucalypt diversity and evolution and its impact on koalas, Dean's arboretum, about an hour south of Adelaide, is where I need to go.

Eucalypts are not all that easy to identify. There are currently around 800 or 900 species and subspecies of 'eucalypts' or gum trees, Dean tells me, variously classified as *Eucalyptus, Corymbia* and *Angophora*, and nearly all of them are native to Australia. Most gum trees are true *Eucalyptus*, but about one hundred are currently classified as *Corymbia* (bloodwoods, ghost gums and spotted gums) alongside a further nine species of *Angophora* nicknamed 'apples', although I cannot see the resemblance.

To northern hemisphere eyes, gum trees often seem distinctly un-treelike – presenting a uniform forest of dull grey-green that does not change from one season to the next: 'nevergreens' rather than ever-greens. They are not compliantly round or conical, nor regular, neat or trim in their growth. Individuals of the same species vary widely in form and shape, depending on their circumstances and environment. It is their variety and individuality, not their supposed uniformity or invariance, that is confounding. Their ability to adapt to the particular conditions they find themselves in bewilders efforts at identification. This is part of the great value of the arboretum Dean has established over the last thirty years. In attempting to grow four 'sibling' trees of every Australian eucalypt species in the one place, under the same conditions, he has created a natural experiment in eucalypt variation and conformity to type.

Each of the four trees that Dean plants for each of the species has been established from seed collected in the wild, from the same 'gum-nut' fruit of a single 'mother' tree. They may have different pollen parents, but all the seedlings are at least half-siblings, planted, raised and grown under identical conditions. And yet, as I walk along the neatly ordered rows of the arboretum, I am struck by the individual variation even between the siblings. In one quartet, one has grown tall and straight, another like a dense shrub, a third with multiple trunks, and the last has fallen sideways before resprouting upwards.

The difference is even greater between species. Most gum leaves have a typically elongated crescent or scythe-like shape, but the species vary wildly. The Western Australian square-fruited mallee, *Eucalyptus tetraptera*, superficially resembles a frangipani or magnolia than a gum tree, with enormous rubbery leaves and large pink flowers. The broombrush mallee, *Eucalyptus angustissima*, looks more like a conifer or she-oak with its fine needle-like foliage and tiny clustered buds along its stems. Like acacias, eucalypts transform their leaves dramatically from juvenile to adult, and periodically revert to the garb of infancy when they regrow after fire or injury.

Identifying species is a nightmare for the uninitiated. It frequently comes down to complex combinations of the pattern in the leaf glands and veins, the arrangement of flower buds, or the shape of their gum-nut fruits. Difficult features to see at the top of a 30-metre tree, where the nearest leaves dangle far out of reach.

Koalas, I assume, have no trouble telling eucalypts apart. Or at least they know what they like. Of the hundreds of species of eucalypts found across Australia, only seventy or so are recognised as koala food trees and, of these, any one individual koala might only eat three or five or ten different species.

Dean and his partner, Annett, have just returned from a field trip to Queensland, collecting samples of leaves from each species of tree suspected to be eaten by koalas from several different locations. Associate Professor Ben Moore, a nutritional ecologist at Western Sydney University, wants to use a new approach to understand which trees koalas feed from – by identifying plant DNA in koalas' faeces, and Dean's ability to accurately identify species is essential.

They have returned with 540 leaf samples from 40–50 species, all the way from the most northern extent of the koala's range near Cairns down to the hinterland of the Gold Coast, Sunshine Coast and Brisbane – regarded as the current hotspot for koala abundance.

'Did you see any koalas while you were there?' I ask.

'Not a single one,' Annett replies.

'We see far more koalas here,' Dean comments, waving his hand towards the park across the road from his house in the foothills of Adelaide. Whatever the outcome of the research, that observation – from someone specifically looking up into the food trees of koalas – is not a good sign.

I realise that if I want to understand koalas, I need to understand the plants that power them. Plants are the building blocks from which animals are made. Plants construct themselves – branch and stem, leaf, flower and wood – from sunlight and water and a few essential elements. When herbivores eat, they break down the plant to reconstruct its components into muscle and bone. The carnivores that feed on these herbivores have an easier task of transforming flesh to flesh – more reassembly than reconstruction.

Eating plants is surprisingly complex. The basic building blocks and frameworks of plants and animals are so different. The bones of vertebrates and the exoskeletons of invertebrates are constructed of separate, specialised structural cells. But in plants, structure is more integral – every cell has a hydrostatic cellular membrane that allows it to stiffen and soften by changing its internal water pressure. This tough but flexible membrane is protected by a cell wall made of cellulose, allowing movement with light and water, but with limited rigidity.

Some plants have generated more rigid structures through the use of lignin. What they lose in flexibility, they gain in strength. Lignin and cellulose create wood, create trees, create forests – vast superstructures, the skyscrapers of the natural world that house thriving teeming arboreal ecosystems, stabilise temperatures, create microclimates, change the weather and alter the global climate.

Without cellulose and lignin, plants may never have evolved beyond layers of green photosynthetic slime, floating in a primordial soup or coating the rocks. Instead, they lifted themselves from the ground, stretched leaves towards the sun, expanded, branched, diversified, proliferated. They moved out of the shallow seas and onto the land, locking moisture inside their cells. They sent out life support capsules to seed the land by air and water, colonising every vacant space with herbs, shrubs, woodlands and forests. And where the plants went, the animals soon followed – consuming the plants as fast as they grew.

Land-dwelling crustaceans followed the plants ashore and evolved into insects. They learnt to chew, suck, bore and grind; how to climb, dig and fly. As fast as the insects evolved to consume the plants (and each other), the plants changed to avoid them, growing faster, taller, thicker, tougher, generating toxins, oils and waxes. The history of life on earth is a struggle between the eater and the eaten, between the plants and the animals, the two great kingdoms into which Aristotle, and later Linnaeus, classified all living things.

But on the sidelines, underneath and in between, invisibly infiltrating everything, was a whole diversity of other tiny organisms, single-celled, colonial, chimeric and complex – five kingdoms of bacteria, archaea, protozoa, chromista and fungi. Sometimes they are essential allies, but often they're devastatingly powerful foes. The hidden and poorly understood activities of some of these organisms underpin many of the great mysteries of evolution.

Plants, animals and microbes – it's a trophic trifecta of evolutionary adaptation that drives so much of life's diversity. Small wonder, then, that it is the intersection of eucalypts, koalas and microbes that lies at the heart of understanding koalas.

While koalas have been shaped by the eucalypts they eat, the eucalypts have in turn been shaped by the peculiarities of Australia's climate and geology.

Australia sits in the middle of a tectonic plate, surrounded by a continental shelf that extends like a geo-strategic buffer, isolating us politically, culturally, biogeographically and geologically. Over recent millennia, Australia has been seismically inactive. The Ring of Fire that surrounds the Pacific also skirts the Indo-Australian plate. It cracks the seabed far to the east and raises vast mountain ranges to the north, across New Guinea and South-East Asia, but leaves Australian soils largely undisturbed, to weather and erode without being renewed by subterranean upheavals. In the absence of significant recent glaciation, mountain-building or volcanic activity, Australian soils are for the most part old: decayed and leached of nutrients like nitrogen, phosphorous and potassium that are vital for plant growth.

Phosphorus and nitrogen are also essential for the animals that feed on plants, as are the proteins created by these elements. The paucity of these nutrients and many others in ancient Australian soils posed a great challenge for Australian plants over their long evolutionary heritage. More recently, over the last 100,000 years, Australian plants have also had to adapt to an increasingly dry and arid climate. The ones that have succeeded in meeting this challenge have thrived and diversified and now dominate 90 per cent of Australia's woodlands and forests – the eucalypts, and the koalas that eat them.

We think of Australia as a land of drought and desert. Both koalas and eucalypts are models of drought adaptation, defined by their remarkable ability to cope with little surface water. Yet the more I look into their evolutionary history, the more I realise that their origins are in a damper, swampier world. Like most Australian plants and animals, koalas evolved in wet forests, not dry. They are descendants of the ancestral 'mesic' or moist forests that once covered much of Australia. Over the last 2 million years, our climate has alternated rapidly between warm, wet conditions and cold, dry ones. With each oscillation the damp forests retreated into isolated pockets, while deserts and grasslands expanded. Each retreat retracted the range of all the species within their isolated forest refugia, separating them from other populations for thousands of years and over countless generations. When warmer conditions finally returned, the forests expanded and reconnected, but many of the species within them no longer recognised one another, could no longer breed with their neighbouring populations. They had become separate species during these interludes of isolation.

These climate fluctuations explain much about the distribution and diversity of Australian plants and animals. They generated the mosaic pattern of Australian forests, compared to the vast swathes of more homogenous forests in the northern hemisphere. They gave rise to the huge range of eucalypt species across Australia and why, with up to 900 different species, there are barely a dozen widespread species across the continent while most are restricted to tiny ranges often with just a handful of species native to any one particular area. Climate fluctuations also explain why those widespread species tend to be associated with permanent waterways, floodplains or swamps, like the river red gum, ghost gums and even the more drought adapted coolabah.

These fluctuations also underpin the patterns of speciation and diversity in Australian birds, reptiles, invertebrates and mammals even when there are no mountain ranges or rivers or lakes or glaciers to isolate populations and create different species. Climate and continental isolation, rather than internal geographical features, have been a dominant driver of Australia's biodiversity.

Yet if the climate-driven fluctuations in the forests have generated so much diversity in eucalypts, why has it not driven similar diversity in the koalas that depend on these trees? How is it that koalas, unlike so many other groups of plants and animals, are only one species, instead of many?

I'm raking the gum leaves away from the woodshed. They collect here, against the shelter of the iron sheeting, piling up in a tangle of twigs, branches and bark that crackle and crunch underfoot. These leaves fall all year round, retaining a constant lacy shade beneath the drooping mature eucalypt leaves that often hang vertically to avoid an excess of heat and water loss.

Everything about these leaves is sturdy and resilient. When I put the soft deciduous fruit tree leaves into the compost they break down rapidly, if not quite as fast as the kitchen scraps, or the soft green waste from my vegetable garden. But the gum leaves take forever to break down. Even when they fall in the dam, they drift slowly to the silty floor and lie there, their long, thin, sickle-shape still visible through the tannin-stained water. On the dry forest floor, they seem to be eternal.

I pick one up and crack the brittle leaf along its central vein. Pungent eucalyptus oil perfumes the air, despite the leaf's dryness, and I can feel the residue of the waxy veneer coating my fingers.

These leaves are not disposable. They are built to last. Australia's low-nutrient soils make it costly for trees to grow leaves, so they do not waste the ones they have. Nor do they let them parch under the fierce Australian sun. These leaves are thick and sturdy, closed off from the outside world, their waxy coating giving them their characteristically grey foliage. They are filled with cellulose, like all leaves, but these ones are also stiffened and strengthened by lignin, the material used to create wood.

If you want to make replicas of oak or maple leaves, you might cut them from the thin pages of a paperback book. If you wanted to make gum leaves, you would have to cut them from the cover. A matt lamination finish would be perfect.

The defences of eucalypt leaves are prodigious. It's not just that they are physically tough; they are also downright toxic. Eucalypts employ a wide range of defensive chemicals to avoid being eaten – the most obvious of which is eucalyptol or cineole, the oil that gives the trees their fresh distinctive scent.

Eucalyptol is a terpene – a group of volatile compounds that frequently make up fragrant essential oils, like menthol, camphor, citral or thymol. Terpenes are part of the self-defence system of trees, having powerful fungicidal, anti-bacterial, healing and insecticidal qualities. They can also suppress the growth of understorey plants that might compete with the tree for resources.

These terpenes give forests their spicy scent. They are released as volatiles that oxidise into aerosol form, sometimes creating nuclei for condensation, bringing rain, brightening clouds and cooling the climate. In Australia's hot, dry seasons, eucalyptol helps fuel the forest fires that burn fiercely across vast swathes of country.

On the generally infertile Australian soils, plants are more likely to use terpenes and phenolics (such as tannins) for defence, instead of other toxins, as they can be created through photosynthesis, rather than using valuable mineral nutrients drawn up from the soil. These toxins work in a variety of ways. Tannins form long chains that tangle and bind with the equally long protein chains, making the protein indigestible. Some phenolics have to be removed by herbivores by combining them with glucuronic acid, which is energetically costly, making the plant matter less nutritious. Terpenes, which eucalypts contain in abundance, are toxic in large doses to most mammals, insects and even bacteria. Some eucalypts contain high levels toxins of formylated phloroglucinol compounds (or FPCs) like sideroxylonal – which triggers nausea in many leaf-eating marsupials. They even contain the precursors of cyanide. The nutritional treasures in these leaves are locked inside a dungeon, within a labyrinth, protected by a fortress – and booby-trapped at every turn. The more I find out about eucalypt leaves, the more astonished I am that anything manages to eat them at all.

I rake the leaves into a pile and set them alight. The flame catches with a sudden flare. Gum leaves burn hot and fast – a perfect tinder of a dry tough plant material and flammable oils. I live in a fire-prone bushland area, and we can't afford to keep this much fuel around the shed.

As the leaves burn, I wonder about how much time or effort it takes for the koalas to digest these sturdy and durable leaves. No other mammals feed quite so exclusively on eucalypt leaves as koalas do. Cellulose and lignin, tannins and toxins – just how do koalas manage to break down these leaves, when even soil bacteria and detritivores seem to struggle with them?

There must be something quite incredible going on in their bellies.

8

You Are What You Eat

'There's a koala up by the driveway,' my husband says. 'Do you want to go look?'

'What's it doing?' I ask. The narrow band of trees that separates us from our neighbour's paddock and the road is a koala highway. They frequently move along it, but rarely linger. Most of the trees are messmate stringybarks and a selection of ornamental eucalypts planted decades ago by a previous owner. Perhaps not the koalas' favourite food.

'It's just sitting there, not doing much,' Mike replies. 'There's some heavy rain coming, so you'd better hurry.'

It's drizzling as we wander up the driveway, and by the time we get to the gate the rain has settled into a heavy patter. I look up, expecting the koala to be hunkered down, head tucked in against its chest with its back turned against the approaching weather. That's what I would be doing if I lived in a tree.

Instead, the koala is clambering around in the tree, reaching for leaves and munching. Looking through the binoculars, between rain-drops, I can see leaves steadily disappearing into the koala's mouth. It moves from branch to branch, pulling the leaves towards its nose, then mouth, like a child in a sweet store. I don't think I've ever seen such an active koala before. The rain clearly doesn't bother it at all.

I remember reading somewhere that koala fur is so thick and waterproof that it was once popular for lining the greatcoats of northern armies in Siberia and fur-trappers in the depths of the Canadian Yukon.

The rain increases its tempo and my binoculars blur and distort under the downpour. We make a dash to the house but, glancing back, I see the koala leap from one branch to another, looking for all the world like it's having a fabulous time.

Eating gum leaves is generally seen as a bad idea. Gum trees were once planted overseas because most mammals don't eat them, not even goats. Even in Australia, western grey kangaroos – which seem to eat even the toughest and prickliest plants in my garden – won't touch them. Swamp wallabies love the soft leaves of young eucalypt saplings in plantations but in natural forests they only form a small part of their broad diet. Greater gliders survive almost exclusively on gum leaves and flowers, brush-tailed possums include the leaves in a mixed diet of flowers, fruit and a variety of a leaves, and ringtail possums will eat gum leaves but are sensitive to some of their toxins. But only the koala is entirely specialised on these apparently unpalatable leaves.

Their choice of food is widely blamed for a lot of the koalas' problems. Gum leaves are tough, indigestible, low in nutrition and, bluntly, poisonous. They seem like a toxic and indigestible cocktail for us, but for koalas they are clearly just perfect.

Koalas are notoriously picky eaters, though. They are famous for 'only' eating eucalyptus leaves. And not just any eucalyptus leaves either. Where they are available, koalas will exclusively feed on manna gums or river red gums, and in other forests they will restrict themselves to just a handful of species, rarely more than ten. And they prefer the leaves of some individual trees over others. And of those leaves, the zookeepers assure me that they generally prefer the tips – the youngest, the freshest and the juiciest.

Establishing which tree species koalas eat is more difficult than you might think. No-one can be completely sure. It's traditional for dietary studies of wild animals to involve hours of collecting droppings, or scats, in the field, and then even more hours pulling them apart under microscopes, painstakingly identifying their contents. I've spent many hours separating out bones, fur and shell fragments from carnivore scats, identifying them down to genera and species. That traditional approach, however, is not quite so easy for plant matter – seeds, pollen or nuts might be readily identifiable, but it's tedious and time-consuming to identify distinctive leaf cuticles in chopped up leaves beyond the broadest grouping – grass, pine, eucalypts or wattles. Today, new molecular techniques are used to identify eucalypt genome markers in scats.

Early studies assumed that the trees in which koalas were found must be the ones they were eating. In a national survey of koalas conducted in the late 1980s, trained observers were asked to record not only where koalas were seen, but also what type of tree they were found in. In South Australia, nearly three-quarters of koalas were seen in manna gums, as well as South Australian blue gums, river red gums and a few in swamp gums. In Victoria, manna gums accounted for almost half of all sightings, followed by a range of species: messmate stringybarks, swamp gums, narrow-leaved peppermints, southern (or Tasmanian) blue gums and river red gums. In New South Wales, the range of trees was quite different and more diverse: Sydney blue gums, tallow wood, blackbutt, grey gum and flooded gum were the most common. In Queensland, forest red gum accounted for almost 40 per cent of sightings. The rest of the sightings were from a wide range of unspecified species. What's more, koalas were sometimes seen sitting in non-eucalypt species.

We can't always assume that the trees koalas sit in are the ones they are eating. They only spend a small amount of time eating, and there may be other reasons for preferring a particular tree: shade, distance from other koalas, safety, even comfort. In hot, northern climates, koalas spend a lot more time in 'shade' trees rather than feed trees.

Koalas do, in fact, eat non-eucalypt species from time to time. They are known to eat wattles, tea tree, she-oaks and paperbark species, even pine needles, although the bulk of their diet is made up of eucalypts. It's not clear, though, how nutritious these non-eucalypt species are for them. Koalas were observed eating pine needles on French Island, where the population is frequently overcrowded and outstrips its preferred food supply. It's not always clear why koalas choose to eat certain leaves and not others, even from the same tree.

The best way to be sure that a koala is eating the right food is to let it choose for itself, ideally from a forest with plenty of different trees. Because it's not just a matter of what they choose to eat, but whether or not they can digest it.

Just how nutritious are eucalypt leaves? Are they so deficient in calories that koalas need to conserve their energy by sleeping for over twenty hours a day? Koalas do sleep more than other herbivores, and they do use less energy – about 74 per cent – than other mammals of a similar size. Is this because they must save energy, or simply because they can?

Most animals that sleep for twenty hours a day are carnivores with highly nutritious diets. Most animals that eat low-quality food, like sheep, deer, elephants and giraffes, barely sleep at all, surviving on little more than two to four hours a day – much of that while standing up. If koala diets are so poor, then shouldn't they be awake and eating more often, rather than sleeping?

When I look more carefully at koala diets, though, they are similar to those of other herbivores of the same body weight. Koalas eat up to half a kilogram of leaves a day, between 2 and 4 per cent of their body weight. Sheep and cattle typically eat 2–3 per cent of their body weight in dry feed every day. Grass is also a pretty poor-quality food, high in abrasive silicates and even lower in nutrients than most leaves. But its cellulose fibre is more digestible than the more lignified fibre of eucalypt leaves – and grass doesn't have anywhere near as many toxins. On balance, it's probably six of one, half a dozen of the other. However it works, gum leaves must be providing just as much nutrition to the koalas as grass and foliage do for sheep and cattle. The question is, how are they doing it?

Now that I think about it, sheep aren't always grazing when they are in their paddock. Like other ruminants, they spend a good part of their day, particularly in the afternoon, chewing their cud – regurgitating the food they have already eaten and rechewing it to break down the tough cellulose fibres in the grasses. Whatever low-quality plant material that these animals are eating, it takes time to break it down into a digestible form.

Koalas are not ruminants and they don't regurgitate or chew a cud. Whatever system koalas use to break down their high-cellulose/high-lignin diet, they seem to be doing it in their sleep.

Plants and animals are engaged in a millennia-old arms race with each other: plants protect their precious nutrients from theft, while animals try to circumvent these safeguards. The tough cellulous and lignin walls of plants allow them to grow into large, complex shapes and structures, creating and exploiting a wide range of niches, but they also make it harder for other creatures to eat them. Plants defend their cells, inside and out, with spikes and spines, gritty silica and calcium, hairs and irritating needles.

In response, herbivore teeth have evolved to rupture these sturdy walls, fragmenting the fibrous plant material, releasing nutrients and increasing the surface area of the food so that enzymes and microbes can digest them further. While carnivores have a range of stabbing, slicing and slashing teeth for eating meat, herbivores tend to have snipping teeth at the front for biting and large grinding molars at the back.

These molars take a wide variety of highly sculpted and characteristic forms. Most of the cud-chewing ruminants – like cows, sheep and deer – have crescent-shaped 'selenodont' teeth to help them grind down their poor-quality food. Selenodont teeth are also found in the distantly related camels, suggesting that similar strong evolutionary pressures are being exerted on both species in relation to processing their food. And, uniquely among the modern marsupials, koalas too have selenodont teeth.

You can tell if a leaf has been chewed by a koala. It looks like someone has attacked it with a pair of dressmaker's pinking shears – the ones that cut a zig-zag pattern. The leaves are cut into tiny particles by this shearing/slicing action – not crushed or ground. I rub my finger across the molars of the koala skull on my desk. The four pointed pyramids that make up each tooth are still razor-sharp, and when I put the upper jaw against the lower they lock precisely into place – a double row of shears slicing with each bite.

I consider how rare it is for us to bite our tongues (and how painful it is when we do) and imagine how much more important the precise control of their chewing must be for koalas.

These almost architectural tooth shapes do not just emerge fully formed in herbivore mouths. Like any blade, they require regular sharpening. Indeed, they are often sculpted into the specific shape needed to fracture the food they process. The sharp edges of the crescent-shaped teeth of cows, sheep and koalas are formed between hard rows of enamel and the softer dentin that has been exposed by abrasion and attrition. As a result, most herbivores grind their teeth, including humans, who tend to do so in their sleep. Perhaps koalas do too – it might explain the mysterious sounds of mastication in the absence of food that scientists have sometimes detected on their recording equipment. Tiny baby guinea pigs even grind their teeth in utero so that they are ready to use as soon as they are born.

I wonder if tiny koala joeys grind their teeth in the pouch. Has anyone looked?

I love the smell of eucalyptus: that pungent camphorous scent that seems to clear your sinuses like a crisp gully breeze blowing through your head. My grandmother regularly dosed me with it when I was a child suffering from asthma. The medicinal value of the eucalypts has long been recognised for respiratory ailments, first by Indigenous Australians and later by European settlers who shared this knowledge around the world, leading to the widespread use of eucalypt oil for all sorts of ailments, from influenza to malaria. When Mary Poppins sang about taking medicine with a spoonful of sugar, I imagined that she was referring to a spoonful of oil, sugar and eucalyptus, the favourite home remedy of post-war Australians, rather than the polio vaccine.

Modern science has clarified and refined our understanding of the benefits and toxicity of eucalypt oil. Eucalyptus has fine medicinal properties for topical application on the skin, but for most of us, and particularly in large quantities, it is toxic for internal consumption. I never gave it to my own children, for all my fond memories of my grandmother.

If you've got the right stomach for it, I can see why a koala might eat it.

Apparently there is a story that eucalyptus is a psychotropic drug and that koalas spend so much time asleep because they are 'stoned' – doped-out drug addicts, sleeping off the effects of the toxic food. It sounds like a story from the '70s.

Should you ever sneak up on a sleeping koala in its tree and wake it up, it is unlikely to behave like it is under the influence of drugs. Like most of us, they are happy to wake up slowly with a few stretches, a sleepy scratch and a yawn – but under threat, they snap to attention as quickly as any animal and are as ready to retaliate with long claws. There is nothing doped out about a disturbed koala.

But the toxins in eucalypts are real. So how do koalas manage to survive on a diet that would kill most other animals?

I come across a blog post that deplores the stupidity of koalas for not even trying to get out of the rain.

'They will just sit in that rain wondering why they get wet until the rain passes,' the post reads.

It's such a primate thing to say. I'm reminded of the old documentaries of Jane Goodall and her chimpanzees in Gombe, huddling under large leaves in the rain, looking truly miserable. I wonder if it's simply a matter of a shared expression, but some of the great apes in particular do seem to dislike rain very much. Perhaps it's because we don't have a thick waterproof pelt. Most mammals, when you think about it, are content, if not happy, to stand or sit in the rain.

But it's not just about koalas not minding the rain. In the south, where the weather is often cold and wet, koalas tend to curl up in the rain and sleep, waking when it's warmer and drier. But in drier warmer climates and seasons, they seem to love the rain. They often wake up to eat when it rains, day or night. Perhaps it is because the leaves are wet – providing an opportunity to drink without leaving the safety of their trees. Or is it something else?

When trees sense that they are being eaten, they launch a chemical retaliation uniquely developed to exploit the sensitivities of their attacker. Very often they know what insect is attacking them by detecting the molecular structure of the saliva. The chemicals they deploy in response – both phenolics and terpinoids – are intended to deter the attacker from eating too many leaves, forcing it to move on before it can do too much damage. What's more, the smell they generate seems to attract predatory birds and insects who feast on the attackers.

But some trees are more precious about their leaves than others. When trees grow on poor soils, they can't afford to produce lots of leaves, so they protect the ones they have by filling them with chemical repellents. Trees on moist and nutrient-rich soils are more blasé. They can afford to grow lots of leaves, so it matters less if they lose a few to herbivores.

Trees are individuals too, adapting to their own particular

circumstances and responding to the changing conditions around them. They are certainly not mere passive recipients of koala grazing.

'Eucalypt forests do not fit our image of a salad bowl,' concludes Associate Professor Steven Cork, after twenty-five years of studying koala digestion. 'In fact, the more we learn, the more they appear like battlefields through which koalas roam, apparently oblivious to the chemical warfare being waged around them and unaware that there is anything tastier than a tough fibrous low-nutrient potentially toxic eucalypt leaf.'

I wonder if the trees can detect koala herbivory and if they launch a chemical attack in response, the way they do for insects. Are these defences influenced by the koalas' daily and seasonal cycles of activity and torpor, wakefulness and sleep? Are the trees' chemical defences lowered at night when grazing insects are less active? Does rain disrupt their ability to detect the saliva of their enemies? And do leaves taste sweeter in the rain or in the darkness of night?

I can't imagine how long it might take to untangle the complexities of these questions. Neither the koalas nor the trees are giving up their secrets easily.

Maybe it's no wonder we remain puzzled by so many aspects of koala behaviour when most of us spend more time indoors than out. The Harvard naturalist Louis Agassiz famously encouraged his students to 'study nature, not books', but frankly, the library is a lot more comfortable than being out in the rain and dark that koalas seem to like so much. Our dislike of the wet and dark means that we mostly only see koalas asleep in the trees and think them lazy and indolent.

Perhaps they think the same of us as they pass beneath bedroom windows late at night and hear the snoring reverberations of our slumber.

9

The Guts of the Problem

My days started early when I worked at the zoo. I rode my pushbike across the parklands, filled with shrieking parrots in the flowering gum trees, and arrived at the side gate of Adelaide Zoo, to begin work an hour or so before the gates opened to the public.

'You're with me this morning,' said Kerrie, one of the keepers on the Australian mammal round, as I parked my bike. 'We need to feed the koalas.'

We collected our trolleys and wheeled them over to the refrigerated room filled with fresh-cut gum branches stacked in drums of water: vases of foliage in a giant florist shop. Most herbivores can be fed with a mix of convenient and nutritious substitutes – commercial pellets, some boiled carrots or a slab of lucerne. But koalas will have none of that. Each day they must be offered a variety of fresh-cut gum branches. Different koalas like different leaves, different trees, different species; sometimes a tree that had provided food for many months would inexplicably be rejected, leaving the keepers and the horticultural department scrambling to source an acceptable alternative.

Captive koalas have been fed many things in the distant past – tea-leaves, bread, tobacco, even whisky. Efforts have been made to wean them onto gum leaf substitutes, pelletised and processed for maximum convenience. In many instances, the koalas obligingly tried this new food, even consumed it enthusiastically, but without gum leaves – the right gum leaves – they eventually sickened and died.

Charlie wasn't very keen on his gum leaves that morning. He kept his face turned away from them, his eyes steady on his keeper. I worried that he might be out of sorts. Animals rarely tell you when they are unwell, and zookeepers have to be closely attuned to anything out of the ordinary. All too often the first sign of illness comes far too late.

'He doesn't seem interested in these,' I called to Kerrie, as Charlie made his way down the tree to the ground.

The keeper looked over and laughed.

'He just wants a cuddle,' she said.

As Kerrie came closer, Charlie picked up speed and ambled towards her before sitting and reaching out for her to pick him up. She scratched his back and head, cooing to him as he blinked slowly and turned his head appreciatively.

'Here you go,' Kerrie said to me. 'You take him.'

If Charlie thought he had been short-changed by this swap from his favourite keeper, he did not complain. He accepted my pats with good grace, sitting comfortably on my hip, holding onto my shoulder. After a while he looked back at his newly replenished tree, as if suddenly noticing the abundance of fresh greenery. He leant out towards it. It was time for me to put him back.

I watched as Charlie reached out to pull the leaves towards his nose, carefully sniffing each one before biting and pulling his selected leaf off at the base. With each chomp the leaf slowly disappeared, like spaghetti sucked up from a plate, or paper in a shredder.

'Time for morning tea,' called Kerrie, as she wound up the hose and loaded the buckets of koala poo on the trolley with the discarded gum branches, brooms, rakes and shovels.

'I'd kill for a coffee,' she said, and we left Charlie to the serious business of breakfast.

Plants are not easy to digest, and some plants are more difficult to eat than others. Soft new shoots might be very tasty, but a lot of animals live on fibrous grasses and leathery leaves whose tough outer coatings are particularly hard to break down and digest. This fibre is made up of long interlinked sugar chains that are glued together by lignin. Mammals cannot produce the right enzymes to break down fibre – but many micro-organisms can.

As a result, animals that eat a lot of 'roughage' have incorporated fermentation tanks into their digestive systems. Carnivore digestive systems are a relatively simple tube, largely comprised of stomach and intestines. Herbivores, on the other hand, have diverse and distinctive digestive systems – from huge multi-chambered stomachs to variously swollen intestinal chambers and 'diverticular' side pockets.

Herbivore digestive systems are filled with a complex ecosystem of bacteria, protists and fungi to help them break down their food. Sometimes these 'gut gardens' are in the stomach, but in other species they are in the intestines. There are a multitude of ways in which different species tackle the difficult task of breaking down the plants that they eat, but broadly they are divided into two specialised groups – the foregut fermenters and the hindgut fermenters.

Foregut fermenters include all of the ruminants: deer, giraffes, sheep, cows and goats, but also some rodents and marsupials, such as kangaroos. They have large and often highly complicated stomachs. Retaining their food in the foregut allows them to use another strategy to maximise digestion: once their food has been partially fermented, they regurgitate a 'cud' and chew it again. To ruminate means to 'chew over again' – the origin of the name 'ruminant' for a large group of hoofed herbivores.

Koalas, however, are hindgut fermenters, along with elephants, horses and rhinoceroses, many rodents and rabbits. Their stomachs are more modest, they do little or no rumination, but they have vast and expansive large intestines. It is this lower part of the gut that does the bulk of their digestive work.

I've never given much thought to what happens to food after I swallow it, to be honest. It's a bit of a black box. But I've noticed that bookshop shelves are filling up with more and more titles on digestion – *Gut*, *The Good Gut*, *Gut Garden*, *The Clever Guts Diet*, *Follow Your Gut* and even *Gut Diary*. We are all listening more carefully to our stomachs nowadays, becoming aware of the impact of our food on everything from blood pressure to mental health. Headaches, bloating, gas, diarrhoea, eczema, autoimmune disease or just generally feeling under the weather have all been linked to intolerances to milk, gluten, food preservatives, colourings, sulphites, caffeine or fructose. We dose ourselves with prebiotics and probiotics, sip vinegar, cut out triggers, reduce sugar/fat/carbohydrates while increasing fibre, vegetables and lean protein.

We cultivate our biodiverse bacterial gut gardens, teeming with 300–500 different species of good and bad micro-organisms, to help us digest our food. We have no idea really what's going on in there, but we change our diets and our lifestyles – or don't – and hope for the best. For most of us, our microbiome is just one of many factors contributing to our overall health. But for koalas, it is a matter of life and death.

So just how bad are eucalypts as a source of food? How do koalas deal with their fibrous, toxic diet?

'Eucalyptus is pretty tough stuff,' Associate Professor Ben Moore tells me. Ben studies how plant chemistry affects the animals that eat them – how animals digest and unpack the nutrients locked up in their food. He's done a lot of work on eucalypts and on koalas.

'Generally speaking, the leaves of woody plants or "browse" are much better quality than most grasses. There's a lot more protein in leaves than in grass. But it depends how accessible those proteins are. Grasses have more fibre, but because it's cellulose fibre, it's easier for foregut fermenters like ruminants to digest. So animals like goats and donkeys can survive on really low-quality food. The lignified cellulose found in eucalypt leaves is much harder to break down and digest.'

'And what about the toxins?' I ask.

'That's the other disadvantage of browsing on leaves,' Ben explains. 'Woody plants also have a lot of tannins that reduce digestibility and more toxins to deal with. Most herbivores try to avoid toxins by foraging more widely and mixing up their diet. But koalas have an unusually high toxin load because they only eat eucalypts, which are highly chemically defended. Those toxins mainly have to be dealt with by the liver.'

The liver is the largest single internal organ we have, typically smooth and simple in overall shape. It's important for all mammals and it's particularly large in some animals, like humans and cows. The koala liver, however, is extraordinarily complicated – highly structured and multi-lobed with a distinctly elongated gall bladder beneath it, producing bile to help digest fats. When they are young, koala livers are dull red, the same as they are in most other animals, but as they grow older dark pigments from the gum leaves accumulate in them, turning the liver dark purple.

'Everything that comes from our digestive system goes straight to the liver before it goes off around the body,' says Ben. 'It's a giant filter that does a first pass at neutralising all the toxins coming in. If you can't do that the toxins will spread around the body where they can do harm. The koala liver has an unusually high concentration of different cytochrome P450 enzymes, which are really good at breaking down a range of chemicals.'

'We think toxins like sideroxylonal damage the gut wall and cause serotonin to be released, which creates nausea. But we're not really sure how they work or how they are dealt with,' he adds.

Ben's doctoral thesis demonstrated that these toxins differ between different individual trees and probably have quite a strong impact in whether herbivores like koalas can eat them. Sideroxylonal seems to be a strong deterrent for the more sensitive species, like ring-tailed possums, while newly discovered toxins like unsubstituted B-ring flavanones (UBFs) occur in trees that koalas don't eat, or prefer less, like stringybarks.

It's not clear when or how koalas acquired their ability to detoxify their potent diet, but it seems to have arisen a very long time ago. A large international team led by geneticists Rebecca Johnston and Kathy Belov recently sequenced the koala genome, and they found that the section of the genome that codes for detoxifying proteins is twice as big in koalas as it is in other mammals, like humans. These genes have a particularly high level of expression in the cells of the liver. Over time, they have been accidentally duplicated – allowing greater flexibility in eating eucalypts. The team also found additional genes for controlling the koala's sense of taste – which might very well help them distinguish between not only different eucalypts, but perhaps the different levels of toxins between species or even individual trees and leaves.

The exceptional ability of the koala's liver to metabolise toxins has some interesting side effects. Koalas are notoriously hard to treat for diseases because their bodies treat antibiotics and anaesthetics like toxins, to be neutralised by their livers. A single dose of an antibiotic used to treat humans would need to be repeated for thirty or forty-five days in order to be as effective in a koala.

Toxins, though, are only part of the koala's problems. Even once the toxins are dealt with, they still have to find a way of unlocking the nutrients locked up in the plant cells. They must break down the fibre and untangle the long chains of proteins from the tannins.

'There are lots of different ways of digesting plant fibre,' Dr Ben Moore tells me. 'Everything has evolved a slightly different way of doing it. But koalas are specialised and they are unique and the way they digest their food is pretty remarkable.'

The first trick koalas use is the way they chew their food, snipping it up into little pieces with those sculptured selenodont pinking shears.

'They are pretty tough leaves and hard to break up into small bits, but the koala dentition does a fairly good job of that,' Ben explains.

'It's not very efficient to focus on digesting the large chunks firstly because they are lignified cellulose. And the smaller particles have a relatively larger surface area, so there is more for the bacteria and enzymes to work on, making them easier to break down.'

Plants go to great lengths to make their proteins hard to break down – they defend them, but they also go on the attack. The front-line soldiers in this war are the tannins – long polymer chains that not only tangle themselves up with the proteins, but also bind with microbial proteins and even with the gut wall proteins that protect the lining of the digestive system. Tannins shield the plant proteins and steal proteins from the microbes and the host animal on the way through. If they eat too many tannins, a herbivore could end with a net deficit in protein instead of a gain.

Some animals counter this danger by lacing their saliva with 'sacrificial' proteins, which lock up the tannins in their food and prevent them from damaging the digestive system. There is no evidence koalas do this, but nor is their saliva just a simple lubricant; it's an integral part of digestion, containing a cocktail of enzymes and other molecules of chemical warfare.

The need for specialist additives continues as the koala passes the dry, dense, chewed-up mass of leaves down the oesophagus and into the stomach. The koala stomach is not particularly large, but it does feature an unusual protruding pink cardiogastric gland. While all mammals have mucus-producing glands that line the surface of the stomach wall, only a few have this distinctive additional gland, located at the entrance to the stomach, which secretes a thick pale-brown fluid into the stomach. Only koalas, wombats, beavers and, strangely, the aquatic manatees and dugongs seem to have this gland. What on earth could connect these extraordinarily disparate and distinctive animals, living in drought-prone treetops, waterlogged forests and estuarine seagrass beds? The only thing I can think of is that they all have high-fibre diets of roots, leaves, wood or seagrasses. Also, none of them needs to drink much water, as they all derive the bulk of it from their food.

Perhaps, with their small stomachs and relatively dry food, koalas need a 'gastric juice booster' to increase their mucus without a commensurate increase in stomach size. In beavers, the cardiogastric gland secretes a particularly thick mucus which collects decomposing cells and fine particulates as they pass through the stomach, slowing down their progress to allow for digestion.

Beyond the stomach lies the intestines. In humans, these long and winding tubes fill most of the abdominal cavity, providing a vast surface area through which to absorb nutrients as food passes slowly through the digestive tract, with the bulk of the digestive work being done in the stomach and small intestine. In koalas, though, digestion takes a very different path.

At the end of the 2- to 3-metre-long small intestine, the chopped-up bolus of leaves reaches a fork in the road. One path leads down the large intestine towards excretion, but the other leads to a large cul-de-sac called the caecum. And in the koala, the caecum is huge, nearly 2 metres long. Koalas have the largest caecum of any mammal, irrespective of body size, bar the strange long-fingered, nocturnal Madagascan aye-aye. Our caecum has shrunk back to a vestigial nub, which Darwin dubbed an 'appendage' or appendix. It's mostly regarded as useless, notable only for occasionally causing death through infection.

For koalas, however, the juncture of the caecum and the large intestine is the nexus for a remarkable separation.

'Once the chopped-up leaves reach the caecum, there is a whole range of different particle sizes from large chunks of lignified

cellulose to much smaller particles, including the liquid solute, containing bacteria,' Ben explains. 'Those large chunks are not going to
be fermented anyway – they just take up space. The smaller particles
have a relatively larger surface area for the bacteria and enzymes to
work on. So it's profitable to hang onto the small particles and get rid
of the large particles. The small particles and the solute also include
the bacteria which break down the proteins for the koala.'

As the larger particles of chewed-up leaves pass through the large
intestine, a regular backwash sweeps the smaller particles backwards
for additional processing. The larger particles pass more quickly
through the remaining digestive tract, in about three to five days.
There is little nutritional value in these structural fibres, and it probably takes too much energy to extract it. Most of the value in the leaves
comes from non-structural carbohydrates and lipids, which the koala's stomach has 'rinsed' into a nutritious soup of partially dissolved
smaller particles. This liquid diet flows into the caecum and here, in
this dead-end backwater, it stays for up to seventeen days, while the
koala's remarkable gut biome does its work.

'So how do they separate the large particles to be excreted from
the "soup" that goes into the caecum?' I ask.

'It hasn't been well studied in koalas, but different species all have
different gut wall morphology for separating their food,' Ben says.
'Voles have channels along the gut wall and the fluids get pushed up
that. Most guts have ridged walls to move stuff along.'

I've never seen a koala caecum, but under a microscope it is variously described as having a 'tall fringe of mixed bacteria tenaciously
adhering to the mucosa,' and 'longitudinal folds of mucous membrane'.
I'm imagining something like a pink forest of sea anemones and soft
corals sweeping back and forth in an ocean of tiny micro-organisms,

filtering out the nutrient-rich sediments, although I suspect it's not nearly so picturesque. No-one really knows what any of these bacteria do, in either the human or the koala digestive systems. Some of them benefit us, some of them might only benefit themselves. Whatever their role, without our complicated gut gardens neither koalas nor humans would be able to digest their food. No matter how they work, we rely on them to keep us healthy and well fed.

It is difficult to study digestion in the lab. Ben tells me they use a bucket full of enzymes to see which nutrients are left in the plant sample after it's been digested, but even identifying the bacteria involved is hard. In the past, identifying species required researchers to grow cultures of the bacteria in the laboratory – but not all the bacteria will grow under these conditions. More recently, researchers have been able to use gene sequencing technology to identify all the different genomes of the bacteria found in different parts of the koala's digestive tract. Using this technique, fifty-four different genera or types of bacteria were found in the caecum, colon and faeces of koalas, which is similar to the diversity found in the better-studied human gut biome. Slowly and steadily, we're getting closer to understanding exactly what they do and which are important for breaking down the particular components of eucalyptus leaves for food.

We've changed our minds about our own appendix too, incidentally. No longer is it a vestigial redundant organ, prone to infection and best removed to avoid complications. With our expanding appreciation for gut biomes, we now realise that the appendix might actually provide a safe house for good bacteria in the event of some kind of illness or purging event. It's like a refugia – a protected place where rare species can repopulate and colonise a damaged and empty landscape.

'Koalas have always been described as specialist feeders,' says Ben, 'and among mammals, they really are remarkably focused on a single group of trees. As a species they actually eat a large number of different trees. And even as individuals they feed on several species – as many as ten or fifteen in some cases, and not all of them are eucalypts.'

We have to stop seeing eucalypts as homogenous. They are hugely variable trees – chemically and nutritionally divergent. For koalas to have spread over such a wide area of Australia, across completely different forests and ecosystems, they have had to adapt their behaviour and digestive processes to counter the challenges posed by the well-protected eucalypts and their chemical arsenal.

This adaptability, however, has created some challenges for koalas in the modern world.

While koalas from one location are all generally able to eat the same kinds of leaves, many koalas that have been moved from another location will simply refuse to eat the leaves that the local koalas eat. For a long time, relocated koalas were a huge struggle to care for. Sometimes they would eat, start to look better, before suddenly deteriorating and dying.

And for a very long time indeed, no-one really understood why, or what to do about it.

IV

A LIFE IN REFLECTION

A grey bundle of fur stirs in the fork of the tree. At first, it appears completely spherical: no head, no face, no ears, no limbs, just a round grey lump. Slowly, like poorly folded origami that will not stay tucked in, a black-tipped foot, then a leg, straightens out on one side of the trunk. A tremor runs across the fur – a series of tiny twitches and shudders – as if in response to the irritation of an invisible insect.

The koala finally unfurls and lifts its head, ears flicking with uncharacteristic agitation. She transforms – with four limbs and a head – swinging swiftly up into the higher branches, coming to rest just before the very edge of the canopy. Her ears are fully outstretched, great fluffy satellite dishes sweeping the heavens, even while her eyes blink against the bright light.

She ignores the quarrelsome chatter of honeyeaters and robins around her, the long screech of a cockatoo rejoining its flock. The fluid warbling of magpies, the tink of bell miners, the sharp crack of whipbirds across the gully, and the gobble of a wattlebird from below all merge into the ambient hiss and sigh of the brittle gum leaves rattling in the wind. A pied currawong peals across the valley, fading into the distance.

She's listening for something else. A deep rumble in the far distance, drawn as if from the back of the throat, up through the belly and expelled with every lungful of air into a deep, wheezing, bellowing sigh. It is faint, like distant thunder. The koala listens until the call fades out and disappears.

It's time to go. She clambers down the broad trunk onto the ground and heads east, in the direction of the call.

10

From Pouch to Piggyback

I'm pruning the fruit trees in the orchard when I step backwards and knock over the metal bucket behind me. It rolls down the slope, rattling loudly and sending startled chickens in all directions.

From up the hill, in the large stringybark trees on the roadside, a low grumble emerges. There's a pause, then another grumble, each one successively louder, like the snores of a bronchial asthmatic.

'Now you've set him off,' says my husband.

'He's a bit early,' I reply. 'It's not even spring yet.'

'You can't be too careful,' he says. 'Doesn't want anyone cutting his lunch.'

The gasping grunts of the male koala expand to a rumbling, bellowing crescendo, then subside into a wheeze. I listen carefully for a response, but I can't hear anything in the distance. Maybe the other males are too far away, or perhaps they also think it is too early to start declaring their presence in the annual mating season.

This bellowing is unexpected from such a compact, furry animal. The technical description for koala's mating call is 'snoring-like inhalations followed by resonant growling expirations'. Some call them 'tree pigs' or unkindly compare their calls to the braying of a donkey or the guttural sound of something stuck in a garbage disposal unit. Others note some similarity in tone with Mongolian throat singing, although without the musicality. It is tuba not piccolo: a rumbling that starts deep in the belly, builds and resonates in the chambers at the back of the throat and emerges with stamina and solidity – like a long-distance runner – to carry over long distances, rather than floating away on the breeze like the fleet twittering of birds.

These mating calls have a rhythmic quality that they share with other species: from the two-tone wheeze-and-wail of a tiny male button quail to the panting roar of an African lion. In the countryside, it's easy to initially mistake a koala's bellow for the start of a chainsaw or even a distant jet flying overhead. Koalas make this mistake too. In spring, the males will frequently rise to these fake challenges, even chasing after puzzled farmers riding quad bikes.

The low frequency of these calls, between 0.5 and 5 kilohertz, transmits much more effectively over an expanse of trees and bushland. They can easily be heard a kilometre away, signalling to other males the occupancy of their ranges and avoiding the need for direct confrontations. It's possible, too, that these calls allow females to seek out males for mating when none are in their area. Females also bellow, although nowhere near as much as the males. And even at a distance, these mating calls are individually distinctive. I guess that's useful when you live so far apart, and you want to make sure you are travelling towards the right individual.

The koala's anatomy has evolved to maximise this bellowing. Their larynx is not high in their throat, like most mammals, but is down around the third or fourth vertebrae. They tilt their heads back to call straight up to the heavens – reshaping their oesophagus and trachea to maximise the sound. Their larynx is attached to a muscle on their sternum. By stretching their necks out, koalas may well be pulling their larynx even lower to deepen their tone.

Because larger males have longer vocal tracts, they bellow at a lower frequency. This means that other koalas can accurately guess the size of their potential opponent by the depth of their call. Many male animals check each other out without directly fighting, thereby reducing the risk of harm. Some strut parallel with each other, gauging who is bigger and more likely to win in a fight. It seems that koalas, rather like lions, are able to judge each other's size by the quality of their bellowing. And female koalas can perhaps judge the quality of their mates.

Kath Handasyde told me about one female who had established a home range some distance from other animals and successfully raised a young joey. At the beginning of the next breeding season, she left her joey at home and travelled 2.5 kilometres outside her normal range. The researchers found her on the border of the home ranges of two male koalas, sitting in a tree with one of them, the other just 100 metres away. After a short stay the female returned to her original area, and when the researchers checked on her a month or so later they found a newborn joey in her pouch.

When male koalas do fight it's fairly brutal, with much swiping of sharp claws, biting and pushing out of trees. They square off on the ground like Greek wrestlers – wide, squat and formidable, completely focused on each other and oblivious to any onlookers, even inquisitive dogs. Although they have very thick fur and skin, it's still pretty fierce.

'Skulls from wild male koalas have a lot more damage on the jaws than females,' Kath told me. 'And they sustain a lot more injuries from falling out of trees. That's all from fighting.'

I can see why it's helpful to bellow, not only to spread animals out in a patchy environment and thereby avoid costly fighting, but also to bring them together when the time is right.

A month or two later and spring has arrived, even though the weather is still cold, wet and blustery. The wattle trees are an avalanche of bright yellow blossoms, the rosellas have occupied their favourite nest box, and the tiny blue wrens are flexing muscles at their reflections on my windowsill. Spring comes early in the southern Australian bush.

I am visiting Cleland Wildlife Park today. It's not far from where I live in the Adelaide Hills, and I often bring overseas visitors here. Like many of the specialist native fauna zoos around the country, it's an ideal place to see Australian wildlife in a semi-naturalistic setting – kangaroos, koalas, emus, wombats, bandicoots, bettongs, dingoes and echidnas. Perched on the edge of the escarpment and surrounded by a conservation park, it offers spectacular views over the city and the ocean beyond and has an abundance of walking trails through the surrounding bushland, which stretches from the ridge top of Mount Lofty, west down the hill face and into the eastern suburbs of Adelaide.

Cleland is now a major breeding and conservation centre for koalas. It is home to one of the largest breeding colonies of disease-free koalas in Australia with a research hub, Koala Life, dedicated to improving our understanding of the biology and ecology of this iconic animal. Like the major city zoos – as well as its interstate counterparts at Healesville in Victoria, Lone Pine and Currumbin in Queensland and the Billabong Koala Park in Sydney – Cleland is part of a collaborative national and international program to ensure a viable captive population, to safeguard the species' future in case of some unforeseen disaster in the wild.

It's not the first time such artificial refugia has been needed to 'rewild' koalas after local near extinction. It's unlikely to be the last.

It's not the science of koala breeding that I'm interested in today, though. I want to talk to one of the koala keepers. Most of us only see koalas once in a while in the wild, if we are lucky. Many scientists who study them do so from tissue samples and DNA sequences. We see koalas as a species, a type, as if they are all one and the same. Few of us have the time or resources to spend weeks or years in the wild, observing their natural behaviour, let alone have the training to interpret what we are seeing. Like ecologists, zookeepers are often uniquely placed to know their animals individually, intimately; they understand their tics and habits, their irritations and annoyances, their routines and quirks. I'm hoping that Ashleigh Hunter, who looks after the main breeding colony at Cleland and cares for the koalas daily, can share a more personal perspective on these animals.

'Enclosure' is the right word for the koala habitat at Cleland. Standard Colorbond fencing encloses pockets of woodland filled with mature trees and native shrubs. It is domesticated bushland – a little more trampled, fertilised and watered version of nature. The ground beneath the shrubs is uncharacteristically bare, the result of Ash clearing away some 10,000 koala droppings every day.

'Two hundred droppings each for fifty animals,' Ash tells me as she sweeps up, changes the water and refreshes the gum branches. 'It keeps me busy.'

It is a lot, I admit. Zookeeping is largely about providing food to go in one end and cleaning up what comes out the other. The koalas' pellets cluster conveniently around their favourite feeding and sleeping trees – dry, tidy piles with a crisp eucalyptus scent. There are much worse animals to clean up after.

Roo is a young male koala who was rescued with his mother and
then raised by her at Cleland. He clambers at the top of one of his tree
stumps, leaping casually from one branch to another.

'They can be pretty acrobatic,' says Ash. 'They don't usually jump
from one tree to another, but they can leap up to 2 metres if they want to.'

Roo descends to a horizontal branch, striding across it before sitting in the middle, a picturesque model of a perfect young koala. He
watches me warily but unconcerned, his ears tracking Ash's every move
as he patiently waits for the latest selection of fresh gum branches.

'Roo's pretty interactive,' Ash says. 'If he's hungry he'll come down
on the ground and even reach up my legs or try to hold onto the feed
pot. The older females next door, Miss Patty and Sophie, would never
do that. They all have their own way of doing things.'

I ask Ash to explain the breeding program here.

'We use a kind of naturalistic approach,' she says. 'We house one
male with three or four females during the breeding season and let
them do their own thing. This prevents males from fighting with each
other. Some zoos use the "love shack" model, where you put females in
with one male when they are in heat for forty-eight hours. But we prefer
to just let them get to know each other and do things in their own time.'

Koala breeding season is not a very peaceful time. The bellowing
of the males is soon joined by growling, grunting and screaming. It's
hard to know what is fighting and what is not. It reminds me of the
description of koala domesticity by May Gibbs in her classic Australian
children's book *Snugglepot and Cuddlepie.*

'Scream after scream filled the night,' she wrote. 'They were fighting, and as they hit each other they screamed with rage.'

I've read other accounts of the koalas' 'robust courtship' in studies
on French Island in Victoria. The males apparently begin pursuing

females in August, but the females aren't interested in mating until September or October and rebuff male attentions until then. There is nothing polite about their courtship. The often-belligerent males chase the females up a tree, where they take refuge on smaller branches that the larger males can't climb. The females bellow, snarl, squeak and scream.

'Are they very aggressive when mating?' I ask Ash.

'Oh yes, lots of biting and swiping,' she replies. 'When the girls aren't interested, they really beat up the male. I'm not sure how bad it is for them, though; their fur is so thick.'

I wonder if the combative mating on French Island, which is famously overcrowded with koalas, was due to the large number of males in a small area. In the civilised confines of Cleland, the male has his ladies to himself. He can afford to await their pleasure.

In theory, koalas have a polygynous mating system – with one male mating with more than one female. For many years, koalas were included in a long list of species – birds, lizards, deer, baboons and seals – where mature males were described as having a small harem of females, 'jealously guarded'. It was a popular interpretation of mating systems at a time when most of the work in animal behaviour had a strongly masculine focus, largely on sexual selection and competition between males. It was assumed that males fought for access to females, while the female koalas passively accepted the attentions of the dominant males. Over the last forty years, the 'feminisation of biology' has, despite some resistance, shifted the focus to mate choice by females, parental care and the importance of foraging and nutritional demands, particularly during lactation. We now look at both sides of the coin – recognising that both sexes make decisions to maximise their own reproductive potential.

Female koalas are, in fact, quite particular about the partners they choose to mate with and have a variety of strategies to ensure that they mate with the males of their choice. They approach certain males and avoid others. Males seem to be slightly less discriminating, ardently bailing up any female koala even when their advances are ferociously rebuffed. These females often take refuge in treetops that can't support the larger males.

The end result, though, seems to be the same, with males typically mating successfully with only one or two females. There is even some evidence that koalas are more likely to mate with females they've mated with before, rather than with new partners, which may also be a function of female choice.

The females begin to show signs of being interested in mating, regardless of whether there are males around, by pacing, flicking their ears constantly and twitching, or having whole-body tremors. They will even bellow of their own accord, or in response to the male's calls, presumably as a way of making contact.

Despite this interest, the females appear not to release their eggs unless there are males around. Unlike humans, for whom ovulation is a regular 28-day cycle, koala fertility is highly irregular but lasts for up to ten days. The release of the egg into the reproductive track is triggered by the act of mating and the presence of semen. This induced ovulation is characteristic of solitary animals where the opportunity to mate arises infrequently. Regular, spontaneous ovulation is more common in social species where potential mates are readily available.

The marsupial reproductive system is not quite as strange as the egg-laying, milk-producing monotremes, but it certainly has some distinctive features. Unlike placental mammals, which have a single vaginal track and uterus attached to two ovaries, marsupials have two uteruses and two vaginal tracts to carry the sperm to the eggs. In between, there is a third vaginal tract through which the young are born.

Two vaginas, logically, require two penises for fertilisation, and indeed, male marsupials have what's called a bifurcated, or double-headed, penis which fits neatly into each of the two vaginas. Copulation is brief – it lasts almost precisely for a minute and a half. According to Sarah Eccleston at the Currumbin Wildlife Sanctuary, the male thrusts exactly forty-two times before ejaculating – a perfect model for a more bawdy Australian version of the 'eats, shoots and leaves' punctuation joke.

The female will not mate again until after her young is born and weaned a year later. Despite the males' vigorous efforts to persuade the females to mate, the females appear to be successful at deciding who they will mate with. There are many unsuccessful copulation attempts and it seems that, in the end, males at most father only two offspring in any one season. Their testes are located above the penis, like those of other marsupials, and are quite small, suggesting that there is very little sperm competition between males and that each female only mates with one male in a season. This makes both male and female koalas very picky about their mates, which might explain their vocal disputes during breeding season. It's important to choose a good mate early and not waste time on bad ones. Males seem to favour younger females because fertility declines with age and, interestingly, they appear to prefer females they have previously mated with over a new female.

From the ground, koala courtship and mating seem noisy, aggressive and possibly unpleasant. But there is clearly a lot more going on than meets the eye. We can't judge other animals by our own strange primate standards – human mating habits tend to combine promiscuity with aggressively guarded parental pair bonds. Perhaps it is not the koalas who have a strange sex life, but us.

Late pregnancy and birth is a fraught and uncomfortable time for humans. We become conspicuously distended, short of not only breath but also patience. Birth can be both protracted and risky, squeezing big-headed babies through narrow hips restricted by our bipedal stance. Placental mammals typically keep their babies inside for as long as possible, but eventually these fragile little creatures must face the dangers of the outside world.

For marsupials, pregnancy and birth are quite different. The labour is barely visible. If you were watching, you might not even know the koala was giving birth. Kangaroos sit still and lick their bellies and urogenital tract, but koalas are already fairly sedentary, so any such behaviour is less obvious.

Wild koalas can be quite sensitive about their pouch being touched when they have young, although Ash tells me her koalas don't object and they can check them regularly for young. The outward and slightly downward opening to their pouch constricts to keep the joey from falling out. The pouch is too small for the mother to be able to clean inside, but just before giving birth, koalas secrete anti-microbial compounds into the pouch which significantly inhibit the growth of harmful bacteria like *E. coli*.

From conception to birth is only thirty-five days. Pregnancy in similar-sized monkeys takes nearly five months. Like all marsupials, infant koalas are called joeys and are born barely bigger than a thumbnail or a kidney bean. When the single koala joey – or, more rarely, twins – is born it consists of little more than a mouth and two large grasping forearms; the place where its hindlimbs will grow is indicated only by tiny buds. Weighing less than 0.5 grams, this neonate can breathe, suck, digest and crawl a short way, but little else. It has no fur, no teeth, no visible ears or eyes.

At this stage, koala joeys have only one task. They must climb out of the birth canal from which they have been ejected, up through the mother's fur to the opening of her pouch, climb inside and attach themselves to one of the two teats located inside. Here they remain for the next three months or more, feeding on dilute but immunologically protective milk.

Joeys at this stage are so firmly attached to the teat that some early naturalists thought marsupial young must have grown there – as if budding from their mothers' belly. And even now, koala rescuers who need to find a new mother for a very young, orphaned joey describe the process of transferring the tiny baby to a new teat and pouch as a joey 'transplant'.

Koala development is a lengthy, slow and surprisingly taxing process. During their growth from neonate 'pinkies' to adulthood, koalas increase their body weight a remarkable 25,000 times over. However, koala milk is low in nutrients. Although their joeys seem to use less energy than other marsupials of the same age, mother koalas still have to increase their food intake by 20–25 per cent during lactation. Growing a baby koala is not easy when you live in an environment, and eat a food, that can only just support a wide-roaming adult. It's hard to know which came first: the koala's poor-quality food giving rise to its low metabolism, or its low metabolism enabling it to exploit a low-grade food. Whatever it was, the end result is a long infancy and lactation.

By seven weeks the pinky, as the hairless young are called, has grown 26 millimetres long – just over an inch. Its face has become more obvious on its enlarged head. Its nose is beginning to pigment and the dark bulge of its eyes and the edges of its ears are becoming apparent.

By thirteen weeks the joey weighs 50 grams. Its eyes open, and a delicate, silky grey-brown layer of fur has begun to cover it, much like the first very fine hair on a human baby's head. At about 150 days its first teeth begin to emerge, the long, pointed incisors at the front of the mouth, used for snipping off leaves. By twenty-six weeks the joey is fully covered in fur, will soon cut its first molar and is ready to start looking around at the outside world.

As it takes its first tentative peeks from between the folds of the mother's fluffy belly, the tiny joey does not look much like a koala. Without the thick furry pelt and fluffy ears, it looks more like a wombat. But even before they have the teeth to properly chew, they will try to put gum leaves in their mouths and munch on them, ineffectively.

It will take seven or eight months before they are ready to leave the pouch, and they won't be independent until they are eleven months old. Until then, they spend their time close to their mothers, retreating to the pouch when small enough, and clinging to her back when they are larger. Even at this age, there are still many perils for a young koala.

One of the advantages of working on the Children's Zoo round was providing day-care for baby animals that had to be hand-reared. Many were young kangaroo joeys rescued from pouches after their mothers were killed by cars, but there were also possums, an occasional wading bird, wild-cat kittens, tiny deer, baby goats and even a young seal. They all needed to be weighed and fed regularly with carefully measured and warmed milk, which had been prepared with species-specific supplements to mimic the milk from their mothers – the strongest for the marine mammals, the weakest for the herbivores. Many babies required toileting too. Unable to urinate or defecate on their own, the tiny animals had to have their nether regions gently massaged until they ejected minuscule droppings or a trickle of urine. Infant stomachs are delicate things. We inspected these deposits carefully – checking for changes in colour or consistency. Diarrhoea was common and often a sign of fatal illness.

The marsupials were particularly challenging charges. Reproducing their mother's warm, moist pouch – nature's own humidicrib – was extremely difficult. Maintaining constant body temperatures for tiny joeys is nearly impossible until they grow fur. The littlest babies were a 24-hour job, and many keepers carried them in pouches on their chests, day and night, until they grew too big and were old enough to be transferred to the Children's Zoo.

It was a huge responsibility, caring for these babies, and we knew that fatalities, even at this age, could happen suddenly and without warning. The transition to solid food was often a particularly risky time.

After one morning feed, one of the kangaroo joeys left a runny green smear on the damp cloth.

'Everything going okay in here?' a senior keeper asked me, leaning over the gate to check on proceedings.

'He's got a bit of diarrhoea,' I said, showing her the cloth.

'Hmm – looks like bub might need a bit of magic poop medicine,' she said, carefully assessing him. 'Come on, little fella, we'll get you good as gold in no time.'

She wandered out to the yard where the adult kangaroos were kept and picked up some roo droppings.

'Here we go,' she said. 'These look fresh and firm.'

She crushed a small portion of the dropping into some milk. The joey drank it enthusiastically.

'That should firm things up,' she said. 'Usually, one dose of adult poo is enough to give him whatever microbes he's missing to get his stomach back in working order.'

Impromptu inoculations like this were common practice in Australian zoos thirty years ago. But we have only recently realised how useful this approach could be in promoting the survival of super-specialists like the koala, who need the right microbes not only to digest plant matter generally, but perhaps also to break down the particular individual species of trees they are feeding on.

All mammals, to a greater or lesser extent, require a complex community of gut bacteria to break down their food. But given that we all emerge from the bacteria-free environment of the womb, where do these bacteria come from and how do we acquire the right combination of species?

The answer is probably not to everyone's taste. The polite version, quickly glossed over on websites and in pregnancy books, is that we acquire it from skin, breastmilk and 'environmental sources', as young babies put anything and everything in their mouths.

Vaginal delivery is far from being a clean and tidy process. Faecal contamination is not only common but necessary for good health. Babies born by caesarean sections lack the key microbes they need for good gut bacteria and a strong immune system, and are more likely to carry less healthy strains of hospital bacteria.

It's even more important for koalas to have the right gut bacteria by the time they are weaned. If they don't, they risk being starved by the very food they eat.

When Keith Minchin from the Adelaide Koala Farm first observed mothers inoculating their babies, in the 1930s, he was worried. He thought that one of the farm's healthy breeding female koalas had terrible diarrhoea. Green slime dripped from her behind all over the ground near the tree. This was not a good sign. Then he noticed the joey.

'The head and forearms of the young koala were protruding from the mother's pouch and its face was covered with a yellowish-green slime. The baby was forcing its nose into the mother's cloaca energetically eating the substance from [her] rectum.'

I ask Dr. Ian Hough about this behaviour. Ian is the research manager of the Koala Life foundation based at Cleland Wildlife Park, and has been a wildlife vet for thirty-seven years.

'Oh yes,' he says. 'They love it. They get it all over themselves. It's messy business.'

This sludgy green 'pap' is nothing like the normal droppings produced by a koala. In fact, it mostly comprises that rich soup of bacteria and nutrients that is usually kept in the caecum. Consuming faeces is not so uncommon. Other animals do this as a means of recycling their food through their system. Rabbits, for example, produce soft mucous-covered droppings from the caecum at night which are immediately eaten, and quite unlike the dry hard pellets they make during the day. Koalas, however, only use this process to inoculate their young with microbes from the caecum.

I suspect this is about as far from the tourism brochure image of a koala as it's possible to get. And yet, it is in all probability the very epitome of what has made the koala so successful. It's a very effective and clever system of dealing with a diet that requires specialist microbial assistance. Nature is not only beautiful and brutal – sometimes it's just plain bizarre.

11

Sociable Loners

I watch a mother with her joey in a nearby tree, and she monitors me just as carefully. Every time I try to get a better view, she swings her body away to keep her joey out of sight. She seems unconcerned, rarely looking in my direction, but she is closely noting my motions all the same, matching my every move with a countermove.

The joey would also like to see what I am doing, and is peeking up over its mother's head and from behind her ears, which twist back and forth to capture any unexpected sound. Like most mothers, this one seems to have eyes in the back of her head. She can tell from the joey's shifting weight what it is doing, and she swings like a perfectly balanced fulcrum, always keeping her own mass and bulk between us, however we each circle her.

I have no desire to mess with a mother koala. I remember slowing down to wait for one with a large joey on her back to cross the road. I watched her cross safely then start climbing a tree on the other side. As I drove on, I checked my side mirror, just in time to see her take an angry swipe at the side of my car with long outstretched claws – threatening, even at a safe distance.

Like most children, koala offspring do not always value their mother's fierce protection. Young joeys are as diverse and cantankerous as any primate toddler. Some of them cling tight, cautious and careful about the world beyond their mother's back. Others are rambunctious and adventurous, constantly testing their boundaries and their mother's patience, precariously grasping the ends of branches just out of reach of her disciplinary grip. An entertaining video on Twitter shows a joey perched on a thin bent sapling, while its mother stretches upright from the ground to pull the branch down. The joey is clearly reluctant to comply with its mother's wishes, and is eventually only dislodged by a swiped paw. But before the mother has a chance to grab her wayward young, it streaks back up the sapling out of her reach again.

Every child has to leave home eventually. I've heard stories about mother koalas biting, swiping and growling at their adolescent offspring, particularly young males, and chasing them down from her tree.

At Cleland Wildlife Park, the enclosures often have multiple animals in them, sometimes mothers and their offspring, sometimes unrelated koalas. I ask Ash what happens when the young koalas reach an age where they would normally leave home.

'Do the mothers show any signs of aggression when they are weaning their young?' I ask. 'Do they favour their daughters or their sons?'

Ash shakes her head.

'No, they just gradually spend more time apart,' she replies. 'We house plenty of mothers with their daughters and there's never any aggression. It's only during the breeding season that things get stroppy.'

I suppose these koalas don't really have anything to fight over – with food and shelter being provided on tap and in abundance.

Once weaned, young koalas generally remain in their mother's home range for up to a year before establishing their own. Daughters do not move far, frequently overlapping or adjoining their mother's range in areas likely to contain the same kinds of trees they are familiar with. Young males, however, are less fortunate. They tend to depart the safety of their mother's range at about two years old. For the next three years, they are nomadic until they are large enough to defend themselves against other males. Transient young males must often be forced to shelter in sub-optimal habitats and make do with undesirable food, to which their guts are poorly adapted to digesting. Mortality for young koalas, males in particular, is high. In unfamiliar territory they are more vulnerable to being attacked by both resident koalas and predators.

Dingoes, owls, raptors, pythons and goannas are all known to hunt and kill koalas. In 1878, Queensland politician Oscar de Satgé described coming across two 'eagle-hawks' or wedge-tailed eagles fighting with, or over, an equally pugnacious koala on the Darling Downs. With a wing-span over 2 metres, wedge-tails – or 'wedgies', as they are commonly known – are Australia's largest eagle. Capable of tackling kangaroos, they are certainly able to prey on young or even adult koalas from the treetops. Large white-bellied sea eagles, found along Australia's coasts, no doubt pose a similar risk, although koalas are unlikely to be the primary food for either eagle. Even smaller birds can be aggressive, if not dangerous, to koalas that come near their nests – from sharp-beaked Australian magpies to pluckily ferocious little willy-wagtails.

But it is possible that Australia's largest owl, the powerful owl, might be a greater threat from the air. This nocturnal forest hunter, similar in size to great horned owls, is found throughout the same south and east coast forests as the koalas. It specialises in hunting the many possums and gliders, systematically clearing out patches across its large territory. I remember one scrutinising me with glowing golden eyes as I came down my driveway one night. Within weeks, the nest boxes in the forest around our house in Victoria were completely empty of the usually abundant sugar gliders and possums. Powerful owls also eat koalas – at least juvenile ones and subadults. One of the first published paintings of a powerful owl, in John Gould's *Birds of Australia*, depicts a young koala in its talons. A study of nesting birds in south-east Queensland witnessed seven cases of koala predation by powerful owls, always of young juveniles and subadults, mostly between June and August, when young koalas emerge from the pouch.

Other predators also find their way into the trees. There is little escape from a 2-metre-long lace monitor lizard or a 3-metre-long carpet python. Both are adept climbers, determined hunters and are commonly found in the same forests as koalas.

Even so, the greatest natural risks to koalas come from the ground, not in the trees. Dingoes are far and away their greatest predators, arriving 3500 years ago and presumably replacing the other large land predator, the thylacines, which disappeared from the Australian mainland about 4000–8000 years ago. Dingoes are especially menacing for females carrying joeys, particularly in more open forests where they cannot travel safely through the canopy. According to some studies, dingoes account for most koala deaths in Queensland, but not in the south, where they have essentially been eradicated since European colonisation.

If a koala survives the perils of its youth and finds a safe and sustainable home range and avoids injuries and illnesses, it can live for quite a long time. It's generally assumed that fifteen years is a fine age for a koala to reach. As they get older, their teeth wear down and they can no longer chop the leaves up into small particles, perhaps limiting their ability to digest their food properly.

And yet there are captive koalas who live much longer than this. Midori turned twenty-four in 2021 and is still active, healthy and much loved at the Awaji Farm Park England Hill in Japan. She took over the title of the oldest-known koala in the world from Sarah who, born in 1978, died at the Lone Pine Sanctuary in Queensland in 2001 at twenty-three years of age. Several other captive koalas have lived for eighteen years.

Life in the wild might seem more dangerous and less predictable, but in fact there have been several accounts of wild koalas living to eighteen years of age as well. Most were female and, remarkably, still raising joeys.

Koalas live in forest neighbourhoods of variable sizes, sometimes overlapping, sometimes adjoining. These are not strictly defended and demarcated exclusive territories, nor are they family groups of related males, females and young. Mapping a koala's home range, by following its activity and movements over weeks and seasons, reveals a tangled cobweb of trails clustered in a particular area, through which they traverse more or less regularly, alone, alongside or over-laying with others. Some animals (like breeding females) are more stable and settled in smaller home ranges shared with successive young, while others (like dispersing juveniles or males) might range over larger, less clearly defined areas.

For koalas, it seems to be environmental factors that have the biggest sway over how they live and breed. The size of their home range, and how much they coincide with others, depends on the quality of the forest. In the semi-arid regions of inland Queensland, where the forests sprawl into dry open woodlands, a single animal might occupy a home range as large as 300 hectares – twice the size of London's Hyde Park and much the same size as New York's Central Park or Beijing's Summer Palace gardens. With less than one animal per square kilometre, these koalas will rarely encounter other koalas. Nonetheless, the presence of joeys and juveniles proves that these are viable breeding populations. Koalas don't travel long distances quickly, so it's likely that bellowing is a particularly useful strategy for helping them to locate partners during the breeding season.

In more productive forests able to support 100–200 animals per square kilometre, female koalas tend to stay in smaller adjoining home ranges, with a male home range overlapping one or two female ranges. In the dense, high-rainfall manna gum forests of Victoria, koalas live in home ranges as small as 1 hectare – the size of an average sports field. With up to 600 koalas per square kilometre, interactions between males are more common, suggesting that competition between males is more intense. In the Adelaide Hills, there are up to 1400 individuals per square kilometre, with some favoured trees being occupied by as many as four or five animals at a time. At such high densities, koala browsing can damage, even defoliate, the trees they prefer to feed on, resulting in starvation.

Koala home ranges create a haphazard patchwork across the forest, centred on preferred feed and shelter trees, concentrated around the rivers and waterways, and expanding as they spread out into drier regions. Those further inland are more likely to cluster around standing water. Annual rainfall seems to be the single biggest predictor of home range, with more rain linked to better tree condition, more free-standing water and smaller home ranges, as well as being associated with the frequency of koalas' preferred tree species and the particular nutritional value of their leaves.

There is much movement between and across these home ranges. Transient koalas travel through occupied areas, perhaps gauging their potential as a new home by the vigour with which the resident demarcates its space, physically or otherwise, or perhaps by the amount of food available. Males tend to be more mobile than females – they are more likely to disperse and to go in search of mates. Most studies find they don't travel far, from an average of 3 kilometres up to 11 kilometres. But this may depend on the quality of their habitat and how broadly we look. Genetic studies have suggested that long-distance dispersals of koalas are actually more common than researchers previously thought – between 16.8 and 20.3 kilometres. In inland Queensland, researchers suspect that koalas can disperse over 50 kilometres in time – far beyond the scope of most study areas. Recently, a couple of young males caused a stir when they appeared in the botanic gardens of the inland town of Emerald – to the delight of the local residents and the annoyance of the vocal and abundant cockatoo population.

In Queensland, the home ranges of males and females do not overlap greatly, no doubt reflecting the fact that they must compete for the same food. But in the richer southern forests, the home range of a male might overlap with several females. The females cluster around patchy pockets of favoured food trees, like manna gums, while the males range across a wider area, probably trading off their less-nutritious food supply with volume.

Exactly how koalas manage these trade-offs – between the nutrition and toxins in the trees they eat, competition with other animals and maintaining mating opportunities – remains unknown. Dr Bill

Ellis has studied koalas in central Queensland for decades, from small dense populations of islands near Mackay to the widely dispersed populations in arid inland Queensland.

'It's a classic travelling salesman conundrum,' he tells me. 'How do you visit the most trees, with the shortest travel time, without visiting the same tree twice?'

If trees respond to koala browsing by increasing the toxins in their leaves, then koalas will need to move regularly around their home ranges and give those trees enough time to 'forget' before revisiting them. It's a problem that has stretched the mental capacities of travelling salesman, campaigning politicians and mathematicians since at least the 1930s. A great many species, from ants to birds to mammals, have evolved different strategies for complex cognitive spatial-mapping tasks – are koalas one of them?

Bill shrugs.

'Maybe,' he says, 'but it could be quite simple. Except for mating males, koalas simply avoid trees that smell of other koalas. And they avoid them whether they are looking for a feed tree to eat, or a shade tree to sleep in.'

It's an appealingly simple explanation, although I'm not sure how it works for koalas in the south, where four or five animals can cluster in the same tree, even when there are plenty of other trees to choose from.

'Well, one thing is for sure about koalas,' says Kath when I ask her about this later, 'every time I study a new population, somewhere I haven't been before, the koalas behave completely differently to what I expected. You just can't predict what they are going to do.'

As I follow Ash around the different enclosures at Cleland, I notice that the animals are not evenly dispersed. More often than not, two koalas will be sitting together – nearby, side by side or 'spooning', with one hugging the other's back, sometimes resting or even sleeping on each other. It happens too much for it to be chance, and it doesn't seem to be related to the proximity of their food.

'Are those two related?' I ask Ash, looking at a huddle of grey fur, with two faces turned to watch us.

'No, I doubt it,' she says. 'Monica and Riley are rescue animals. They came in at different times. They just really like hanging out together. Sometimes Deputy joins in their group hugs, but mostly he keeps his distance.'

We move on to another enclosure.

'These two are interesting,' Ash says. 'Cleo and Vicki were brought in together from Kangaroo Island after the bushfires and were very attached to each other. And they've stayed really close. At some point earlier in their lives they'd been tagged, and their tags are sequential numbers, so I guess they must have been caught together then as well. I reckon they've always stuck close together.'

I'm surprised by this unexpectedly social behaviour in an animal that is known for being solitary. I'd always thought that koalas were fairly aggressive to each other.

'That's only in the breeding season. The males are pretty stroppy then,' says Ash. 'The females can get pretty agitated then too and it's dangerous for the joeys – they sometimes get knocked off.'

I ask Kath Handasyde about her observations of wild koalas. In the sometimes-overcrowded remnants of the Victorian forests, you might expect mothers to move their young on more forcefully.

'They do. Once the young ones get to about 2 or 3 kilos, the mothers will give them a bit of a box to stop them hanging around too much,' she says. 'If they've got another little one coming on, they can't afford to be carrying around a great big joey any more.'

'I think it's hormonal,' Kath continues. 'We had one female who was given a contraceptive when she still had a joey on her back. That boy just got bigger and bigger and she never pushed him off. She was only 6 kilos and he was 4 kilos! They have an amazingly strong maternal instinct. Wild adult females without a joey will sometimes come to the ground and pick up a small baby koala squeaking in distress.'

I wonder about this kind of incipient sociality. Plenty of animals are downright antisocial – hostile, pugnacious, murderous to their own kind, even when resources are not in short supply. It's hardwired into their nature. Common wombats are notorious for being cute, cuddly and affectionate as infants, but downright lethal as adults.

'When they start to mature and hit puberty,' one wildlife officer said about wombats, 'they just hate everybody and everything.'

Koalas, though, appear to be flexible about their sociability – happy to share their space, and even enjoying each other's company, when resources are abundant, but more than willing to go it alone when food is in short supply and it's each koala for themselves. Southern koalas are said to be more sociable than northern koalas, but even here they live contentedly enough in high densities in captivity. Food quality seems to be the key to their apparently solitary natures. Such behavioural plasticity has successfully adapted them to significant environmental uncertainty. Perhaps it is another reason they have coped so well during the fluctuating climate and forests of the last few million years.

We're inclined to judge koalas harshly for their apparent lack of sociability, as if being happy with your own company is a shortcoming. Humans are extreme outliers on the scale of primate and even mammalian sociability. We are obligingly social creatures, living in vast, complex, multi-level aggregations that are rivalled in their intricacies and spread only by insects. We are so determinedly social that we even adopt non-humans into our social networks or invest inanimate objects with human characteristics. We see the world through a brain wired for sociality and interactions with others. As a result, we struggle to understand or accept others, human or animal, whose sociality does not conform to our own.

We define animals as presocial, subsocial or sometimes parasocial – as if there is some kind of evolutionary hierarchy from primitive solitary creatures to highly evolved, complex social systems like our own. But solitary is not the opposite of social; it's the opposite of gregarious. Sociability is a spectrum, not a binary condition, and many species that are usually regarded as solitary – like red foxes or feral cats or badgers – sometimes form communities and exhibit amicable social interactions.

Perhaps we are just oblivious to interactions that are different from our own, or occur outside our limited range of perceptions – such as the constantly refreshing network of chemical signals that lace the landscape, or long-distance auditory communication. The world of blue whales and humpbacks is shaped by long-distance communication over many miles of ocean. The world of many terrestrial mammals is a complex web of chemical signals, mapping the age, activity and interests of individuals who share their ranges. Are you really alone if you are in constant communication with someone else?

There are costs and benefits to sociality. Individuals in social groups might gain more reliable access to mates, shared parenting of offspring, better food-gathering strategies, collaborative shelter building or defence and protection against predators. But living in groups also comes at a price – it intensifies competition for food and shelter, attracts predators and significantly increases disease transmission and interspecies aggression. To put it simply, in many species, competition for food drives individuals apart, while sex brings them together – at least for a short while.

Koalas don't show any obvious signs of social bonding, like grooming each other, but I'm not sure that they are particularly solitary either. When they do meet others (koalas, humans or other animals) they greet them with gentle nose bumps. Other than during the breeding season, they show less aggression towards each other than cats or dogs do on first meeting. Koalas are clearly a naturally dispersed species, probably because of food supply. But if you mapped their activity over a certain area, I wonder if you would find they are more or less likely to interact with one another than they would by pure chance? A bit of both, I suspect, depending on the individual, which is itself an interesting reflection on personal associations – what we might term friendships.

Everything about koalas tells us that food is a limiting factor. Despite the vast coverage of eucalypt forests, finding enough of the right food, the best food, the least-toxic and most-nutritious food – sufficient not only to survive but to breed and raise young to adulthood – is a constant struggle. Small wonder they typically spread themselves as out across the forests, maximising their chances of success in an erratic and unpredictable climate. But this does not mean they are unsociable. Given an abundance of food (and without the fired-up hormones of the breeding season), they seem entirely amicable creatures for whom the touch and smell and warmth of their own kind is as comforting as it is for us.

There is no reason for koalas to associate with animals that are not koalas. Aside from the microbial symbionts that all animals are in constant contact with, I can't think of any benefit a koala would gain from cross-species interactions. They can't offer koalas any protection from predation, or any food or shelter. For the most part, other animals can only present a risk to koalas. So why is it, then, that from time to time koalas seek out the company of creatures not of their own kind?

A friend who raises horses noticed some of her young fillies carefully and cautiously investigating something in a tree in a corner of the paddock. They stretched their noses out towards the low foliage, sniffing, ears forward.

My friend walked around the paddock to see what they were looking at. Young fillies are not always the most sensible of creatures. They bolt and startle at the slightest disturbance, sometimes tangling themselves in fences. She stood ready to intervene and calm them down. Something moved in the vegetation, and she noticed a koala climbing down the tree. It stopped at head height to the horses and slowly turned around, looking at them. The more adventurous of the fillies stretched out her neck, and the koala reached out towards her face. Its long claws lengthened against the soft, delicate skin of the filly's cheek. My friend held her breath. The filly snorted in surprise, huffing horsey breath on the koala, but the koala did not move. The filly leant back in, and the koala seemed to stroke her face before turning back to the tree, and both animals went their separate ways.

We rarely witness such interactions in the wild. It's hard enough to see koalas on their own, let alone with other animals, undisturbed by our presence.

As I wander into the university campus on the edge of the city, I notice a little grey lump in a tree. Koalas are not uncommon here in the tall gum trees and pockets of remnant bushland. But this koala is sitting at eye height in a small deciduous tree, surrounded by buildings and concrete paving, in the middle of a busy thoroughfare with students rushing past on their way to lectures. Security guards have put a fence of orange bunting and a large bowl of water at the base of the tree. Passing students politely line up to take selfies with the koala behind them.

The koala resolutely shuts its eyes and sleeps. It seems strangely indifferent to the presence of all these humans. One of the distinctive things about koalas is that they frequently don't behave like a typical prey animal. Prey animals, even after long domestication, often remain hypervigilant, wary and quick to take flight or hide. But while koalas on the ground will show signs of anxiety and a desire to get away, they are rarely flighty or aggressive unless cornered. And when koalas are in trees, they show few signs of fear. They could be sitting in a coffee shop, glancing up to watch a stranger walk past their table, before turning back to their newspaper and café latte.

12

When It's Smart to Be Slow

The koala was clinging to an old tree stag while stranded in the Murray River, on the border between New South Wales and Victoria. A team of students from La Trobe University noticed its predicament as they were paddling by in canoes.

'It almost looked as though he was sussing out if he could jump into the canoe,' one of the students reported later.

The koala could have swum ashore if it had wanted to – it was close enough, and koalas are not particularly bothered by rain or water. They are capable, if not elegant, swimmers who launch themselves into rivers and swim with an effective doggy paddle to the other side.

If a boat is offered, however, they will readily accept the more comfortable mode of transport. They have been known to haul themselves aboard passing canoes – content to take a free ride to the other side, without showing any concern about where they might be taken.

This koala opted for the easy option. Standing in the knee-deep water, the students spun one end of the canoe towards the tree, where the koala was waiting on a low stump for transport. As the boat touched the tree, the koala immediately clambered on board. The students slowly turned the boat around, keeping their distance from the animal, until the bow nudged the bank. As soon as the boat touched the ground, the koala climbed into the bow before leaping out and strolling off into the trees.

It's an indisputably cute video. Both the koala and the students presumably parted company well pleased with the outcome, but I wonder what the koala was thinking – how it was thinking – about that situation. If you've ever had to rescue a pet from an awkward place – a cat up a tree, a dog stuck in a drain or a horse trapped in a fence – you will know that they very rarely show any inkling that your actions might assist them, let alone co-operate with you. And yet this koala seemed to do both.

I send a link to the video to Mike Corballis, a professor of psychology in New Zealand, who has done a lot of work on foresight and the capacity of animals to 'time travel mentally'. Humans regularly do this – we spend much of our life thinking about what happened in the past and planning for what might happen in the future. Not to mention imagining things that might never happen at all. We are constantly rehearsing scenarios in our minds, revising and refining our responses to interactions, events and conflicts, so much so that an entire 'mindfulness' industry has sprouted to help us stop our whirlwind mental activity and focus on living in the moment.

You'd think that the calm, chilled-out koalas would be the perfect model for living in the moment, but what if they also predict what is going to happen next, based on what's happened in the past, and make plans for the future? The koala in the canoe certainly seemed to do this.

'The koala example perhaps includes problem-solving as well as an element of future thinking,' Mike says. 'It would surely be interesting to do some more work with them.'

The koala wanted to move to a different tree but didn't seem to want to get wet. It saw a means of achieving that goal (the canoe drifting past) and anticipated the possibility that the canoe would come close enough to be used as a bridge, just as the koala might use a floating log. Once on board, it anticipated that the canoe would get near enough to the shore for it to hop off.

It's not clear from the video whether the koala understood the role of the humans in this activity, but it certainly wasn't disturbed by them either. The frequency with which koalas approach humans when in need of assistance suggests that they have some appreciation that humans can provide solutions to problems they are not able to solve themselves. Aside from domestic animals – who recognise that humans can open doors, supply food and perform other simple tasks for them – very few wild animals seem aware of the potential of humans to be useful. And those that do realise this tend to be smart – some of the birds, some dolphins and killer whales, and other primates. But nobody has ever claimed that koalas are smart. Far from it. They are widely regarded as being pretty stupid.

'I'm sure we underestimate animal cognition, partly because we need to believe humans are vastly superior, and partly because we have language and can tell of our plans whereas animals can't,' says Mike. But just because animals don't have language doesn't mean they lack the mental capacity that underlies our evolution of complex language.

We need to stop looking for reflections of ourselves in other animals. There's more than one way to be 'smart'. And accepting a lift from those students to get across the river was, however you look at it, a smart move indeed.

'Marsupials are notably less intelligent than placental mammals, partly because of their simpler brains,' states the *Encyclopaedia Britannica*, in sweeping imperial judgement. It's a widespread belief that has led to many peculiar assumptions about koalas, their ecology and the likelihood of their survival.

In the evolutionary race to supremacy, koalas are regularly pitched as having made poor choices. Like pandas, they are regarded as cute but dumb – soon to be relegated to the growing pile of evolutionary failures, destined for extinction. They are described as slow, stupid and often considered incapable of change. Their diet is often described as so low in nutrients and toxic that it almost poisons them and prevents them from being as active, or as smart, as other animals. If all these beliefs were true, it's a wonder they aren't extinct already.

When I complain to a friend about the negativity around koalas, he looks puzzled.

'Well, they are stupid, aren't they?' he says. 'Isn't that what you get from eating toxic gum leaves?'

The marsupial brain is indeed quite different from that in eutherians, or placental mammals. For one thing, it lacks a corpus callosum, the super connector of bundled fibres that link the brain's left hemisphere to the right hemisphere. Like interstate electricity connectors, this highway is probably more of an equaliser than a one-directional transfer – smoothing the overall transfer of information between the hemispheres, and perhaps allowing one side to take over if the other fails to function. Brains, though, have more than one way of doing the same thing. What the marsupials lack in a corpus callosum they make up for with an anterior commissure, a similar information superhighway that connects the two hemispheres of the brain.

Marsupial brains are also smooth. Mammal brains are characterised by having a 'second' brain – a neocortex that overlays the old structures we share with reptiles that regulate movement, sensory inputs, body functions, instincts and simple stimulus-responses. The neocortex is our rational, conscious brain. It performs many of the same functions as the old brain, but processes information differently. Rather than using instinct, the neocortex is capable of more complex responses to changes in the environment by learning, interacting and making more intricate interpretations of the world. We attribute much of our intelligence to our overly large neocortex while denigrating the cognitive capabilities of animals without one. Whether this is true or not is unclear.

Brains are remarkably flexible organs. They need as much space as they can get, but are constrained by sensory organs in the skull – eyes, tongues, eardrums and others – as well as teeth.

Associate Professor Vera Weisbecker is an evolutionary biologist who heads up the Morphological Evo-Devo Lab at Flinders University. She came to Australia on an exchange from Germany as a student and was fascinated by the country's remarkable, and under-studied, marsupials. Twenty years later, she is a local and world expert on marsupial brains.

'They're hugely undervalued in science,' she says. 'The trouble is that most researchers live in the northern hemisphere, where there is only one species of marsupial – the Virginia opossum. Most of the marsupials live in the southern hemisphere, in South America, and more particularly in Australia, but there are not as many researchers to study them here.'

Vera is convinced there is much to learn from marsupials.

'Firstly, they are a completely different line of mammalian evolution,' she explains. 'They diverged from the other mammals a long time ago and have evolved separately ever since. And they are also very diverse in shape, form, diet and locomotion – carnivores, herbivores, ant-, nectar-, leaf-specialists, bipeds, quadrupeds, gliders and climbers. It gives us a huge range of species, parallel to the eutherian mammals, to study and understand what underlies the different adaptations they have.'

Vera and her colleagues have investigated the different sizes and shapes of Australian marsupial brains. Using the skulls of both living and extinct species, they have created endocasts of the brains – imprints of the inside of their heads. In most mammals, the brain is pressed hard against the skull and squeezed into every space possible. In the past, measuring the size of the brain was done by filling the skull cavity with tiny glass beads and then weighing it. Now the skulls are 3D scanned and the brain shapes can be re-created in intricate detail.

'So are marsupial brains smaller than the brains of all the other mammals, the eutherians?' I ask.

Vera pushes some graphs across the table – clusters of scatter plots with different coloured lines fitted to them, indicating the relationship between brain size and body size for hundreds of species, classified into groups.

'If you look at the lines comparing marsupials versus the eutherians, they follow pretty much the same slope,' she says. 'On average, a marsupial has much the same brain size as a eutherian of the same size.'

'What about these dots that are way above or way below the line?' I ask.

'Let's look at the groups those outliers belong to,' says Vera, moving to a different graph. 'This cluster up the top are the primates. Primates as a group do tend to have larger brains for their size. So do cetaceans.

But sometimes that average is influenced by an outlier. Humans, all the hominids, are really unusual – they have particularly large brains for their body size. They are bringing up the average.'

'Are there any particular outliers among the marsupials?' I ask.

Vera laughs.

'Well, there is one that sits pretty low,' she says. 'Definitely below average on the brain stakes – and it's the Virginia opossum. So I think this is perhaps why northern hemisphere researchers assume that marsupials are dumb. Because they are working with the one species that doesn't have a very big brain.'

'And what about koalas?' I ask. 'Where do they sit on the graph?'

'Let's have a look,' she says, turning to her computer monitor. 'We'll have to hunt for that one. I need to go back to the code and turn on all the labels. It's going to be messy.'

I wait while Vera alters the program and re-runs the graph. The screen suddenly fills with hundreds of species names layered thickly over the top of each other.

'Now, it should be around about here,' Vera says, expanding the screen so that the words start to separate out slightly. 'Ah yes – here it is, I can just make out *Phascolarctos*. Pretty much right on the line – completely average for a marsupial of that size, and completely average for a eutherian mammal of that size.'

It's neither in the top 10 per cent nor the bottom 10 per cent for mammals. There's just nothing out of the ordinary about it. Koalas have a completely average-sized brain for an average-sized mammal.

'There is that argument, though, that koala's brains don't fill the capacity of their skull,' I comment. 'That they only take up 60 per cent of their brain case – which is much less room than any other animal's brain.'

Vera shakes her head.

'There is a little bit of variation in how tightly packed brains are, but not that much. Body evolution isn't wasteful. Why would an animal build a big empty skull it had no use for?'

It turns out that most of the early studies used koala brains that had been preserved, but pickled brains often shrink or dehydrate over time. In addition, brains are often highly suffused with blood while alive, so in death their volume may not accurately reflect their size when functioning. Both of these factors likely led anatomists to think that koalas' brains rattled about in their skulls, floating in liquid. In fact, the amount of fluid surrounding a living koala's brain is much the same as that around the brains of most other mammals.

A more recent study used magnetic resonance imaging to scan the size of living koalas. Rather than a cranial capacity of 60 per cent, this study found that koala brains filled 80–90 per cent of the cranium – just as they do in humans and other mammals.

We really need to radically rethink our common assumptions about the size of koala brains and how they work.

Even if koala brains were smaller than average, it wouldn't necessarily mean that the animals are stupid. Brain size is just too 'noisy', Vera says, to accurately predict mammalian cognition.

'It doesn't reflect the brain infrastructure very well,' she explains. 'Mammal brains differ greatly in their cell density and connectivity, and in any case there is little connection between cognitive performance and brain size or structure either across species or within species.'

Human brain size does not correlate with intelligence. Einstein's brain was significantly smaller than average, sending scientists scrambling for significant differences in his parietal lobes and corpus callosum, or the existence of rare knobs and grooves, to explain his extraordinary intelligence. The relationship between brain structure and function is complicated and only just beginning to be understood. Intelligence may not be a simple matter of how many interconnected neurons you have, but how well those connections are made, pruned and shaped by experience. Brain wiring may be more about the useless connections we lose with age than the valuable ones we strengthen.

Some birds are capable of complex problem-solving and formidable feats of memory, and have mastered tool use and language for their own purposes – rivalling the much-vaunted skills of many big-brained primates and cetaceans. And yet their brains not only don't have a neocortex, but are much smaller and smoother than those of mammals. Flight does not allow birds to develop big, heavy brains, so they have developed small, efficient ones instead. It is not necessarily how much you've got that counts, but how you use it.

Humans are a bit obsessed with brain size – with anything, actually, that we think separates us from other animals, such as tool use, language and sociality. We're a bit touchy, really, about our relationship with the natural world, our place in it. We prefer to consider ourselves different, separated, superior, better.

We admire animals that share traits or habits with us: the prodigious spatial skills of octopuses, the family life of socially bonded birds, the complex communication of cetaceans. But intelligence that does not look like our own, or that results in behaviour or choices different from our own, we don't always recognise or even notice. We think animals are smart when they make choices we would make, even when those choices are dictated by evolutionary selection or instinct, rather than thinking. 'Intelligence' is the ability to make advantageous decisions in a changing and variable world, to solve problems, to adapt behaviourally to shifting circumstances. Some species benefit from being able to do this. Other species, like many sharks or crocodiles, have adopted a strategy that has allowed them to survive unchanged over millennia of changing conditions. Being smart is not always the best strategy.

Dr Denise Herzing suggests that we should use more objective methods to assess non-human intelligence, including measuring the complexity of brain structure, communication signals, individual personalities, social arrangements and interspecies interactions. Ultimately, I wonder if animal intelligence isn't more about behavioural flexibility – the ability to adapt and respond to changing circumstances within the course of an individual's lifetime. This adaptability is even more important than genetic variation for a species' survival – particularly in an environment that is changing as fast as it currently is.

Perhaps we'd be better off spending less time ranking animals on a scale where we are always at the top, and considering them by their own merits and capabilities – in terms of how they live and what makes them successful at what they do. We might have a greater chance of learning something from them that way.

I'm still thinking about the koala that hitched a ride with the students on the River Murray. Like most wild animals, koalas prefer to avoid coming too close to humans. They typically move away, swing behind a tree trunk or simply look the other way. But not always. On rare occasions, koalas tolerate or even seek out human company. They come down from their trees and solicit aid, or simply appear to satisfy their curiosity. It is often younger animals that exhibit this curiosity – who touch noses with people or reach out to them. Sometimes they just seem to want company, which seems odd for an otherwise solitary animal.

In many of these cases, the koala wants something – water or a free ride or safety. They are not the only animals to approach humans for assistance, especially in an emergency, but for others it is rare. Animals do coincidentally use humans to protect themselves, such as a penguin or a seal seeking refuge on a passing boat to escape hunting killer whales, or an injured kangaroo sheltering near a house. Nor do koalas passively accept aid, like a whale that allows rescuers to cut it free from tangled netting and lines. In these cases, the animal tolerates our presence as being a lower risk than the alternative. But these koalas are not avoiding a greater risk; the odds are not so immediately dire. In some cases, the koala might be ill or severely dehydrated. But even so, it is unusual for other animals to actively seek out humans when they are sick. Whatever the koala's health, it suggests an anticipatory awareness of the situation – a capacity to predict future actions and benefits to themselves.

One of my friends once recalled a strange scratching at her front door. When she investigated, she found a koala looking through the glass, apparently trying to get in. Koalas, like a lot of animals, find glass confusing. It's either an invisible impediment that they unsuccessfully try to get through, or it presents the reflection of trees or an unwelcome rival. My friend opened the door and put some water out for the koala as it sat on her front step, apparently unsure of what to do next. When she returned sometime later, the koala was gone.

Was the koala who climbed into the farmer's air-conditioned car, while the farmer was in the vineyard, wanting to enjoy the cool on a hot day? Or was the car simply an interesting obstacle to investigate that happened to appear in her path? It's difficult to know, but even in cars, glass is a problem. It's not easy for anyone to work out how to get around an unexpected sheet of invisible nothingness. What is it that a koala sees when it approaches a window, a human or a building?

I am not entirely sure what it is that makes koalas approach humans when they are in need. Or what it is they perceive when they reach out to bump noses with you. But when a koala does request help, it does so in a way that is intrinsically appealing to humans. Their forward-facing eyes, round face and attentive expressions clearly trigger the facial template that humans are programmed to respond to and read for social cues. Dr Jess Taubert is a cognitive neuroscientist at the University of Queensland who has worked with a range of species on functions like facial recognition, including at the Yerkes National Primate Research Center in the United States. She tells me that people, especially children and those with affective disorders, often respond more strongly to animal faces than to humans.

'My intuition is that animal faces have easier signals to read than adult human faces because we don't always smile when we are happy or stare at what we are attending too,' Jess says. 'Folks with baby faces are rated as more warm, naive, kind and trustworthy and koalas might also benefit from those biases.'

Jess is neither sentimental about koalas nor immune to their charms. She tells a story about being bitten by a koala she was carrying for visitors to photograph when she worked in a wildlife park.

'I knew something was different from the moment I picked him up. I should have just put him down,' she relates. 'He was usually very sweet and patient, but after one or two photos he just chomped down on my shoulder. I had to back away quickly off the exhibit before anyone saw what had happened.'

'He wasn't the only animal to bite me when I worked in zoos,' Jess says, 'but he was the cutest and I instantly forgave him.'

It's not just their faces that make koalas cute. It is also their tendency to lift their arms towards human rescuers when on the ground. It is the action of a tree-climber, an arboreal animal that carries its young and has arms free to lift. As apes, we humans share this instinctive response with koalas. Our infants cling to us, just as the infants of monkeys grip their mother's fur as they ride through the trees. We may have adapted to become fleet-footed, savannah-dwelling creatures, but our infancy betrays our origins. We carry our young like tree-dwellers. Newborn babies grip fingers and objects within reach in a vestigial instinct derived from our primate ancestry, but shared with many arboreal creatures, including marsupials like the koala.

Perhaps when koalas reach up to humans, they are seeking an escape, the tallest object to climb. And when we see them lift their arms, we respond by picking them up. Where they see a tree, we see an infant asking for help. Perhaps we are both victims of our own pre-programmed instincts.

A koala is asleep in one of the trees by the road. I go and check on it a couple of times, but it doesn't move. It is still asleep the next day, but is now on a different branch in the same tree. It must have moved at some point. I just didn't notice it because I was asleep. I think about doing a behavioural activity survey where I check on it every half an hour and record its behaviour, but I decide against it. I'm meant to be writing a book, not doing a zoology paper, and besides – koalas don't do very much, do they? I go back to my desk, where I occupy myself for hours every day in front of my computer. I wonder what my own activity cycle would look like. Long stretches of 'nothing' at my desk, broken by brief forays into the kitchen to eat and perhaps an occasional walk outside.

Then another period of sitting on the couch, and a pronounced period of complete inactivity overnight. I look at the dog, asleep in her basket, and the cat curled up on my bed, and I envy them their relaxed lives. Doing nothing, doing something – it's all relative, isn't it?

It occurs to me that koalas sleep all day because they can, not because they have to. It's certainly not because they are stoned or lack the wits to do anything more interesting with their time. They probably sleep up to 80 per cent of their time, just as cats and dogs do, because they have everything they need in terms of food, shelter and safety. Animals that stay awake all the time do so because they have no choice – because they must move constantly for food (like hummingbirds or pygmy shrews), to fly (like oceanic migrating birds) or swim (like whales), or to maintain constant vigilance for predators (like deer and sheep).

Far from being trapped in some kind of maladaptation, koalas have been set free by their remarkable diet from the anxieties and challenges that trouble so many other species. Once they have found a suitable area, koalas have no need to search for food. They only have to stretch out a hand and pluck it from the tree in front of them, like an emperor plucking grapes from a golden bowl. They have no need for the constant vigilance required by herbivores of African, Asian or American plains. They have few arboreal predators to hide from and their best defence from hunters on the ground is to stay still and quiet and pass unnoticed – even sleeping while they do so. Even their social system requires minimal engagement. They signal their occupation with their scent and respect each other's presence, with almost no contact required. Mating season is the only time that requires any effort, and even then they keep things simple.

All in all, it seems like a pretty good life to me.

Far from lacking cognitive skills, koalas almost certainly need excellent spatial mapping skills to help them navigate their complex arboreal environments. They might have temporal and spatial maps of food availability within their ranges, depending on leaf species, seasonal growth and their own nutritional needs or capacities. They are certainly keenly aware of the movements of others in their environment, and they seem to be able to anticipate or predict their paths through the trees. They recognise and remember individuals (humans and other koalas) over a long time span.

'They have very distinct personalities,' says Kath Handasyde of the wild koalas she has studied. 'Some of them are lovely. Petal was always the most docile and calm animal, while Sonia always tried to bite me. Some of them are bold and some of them are stayers. Some just want to sit on you and others don't. I guess it's good to have a mix in a population.'

And she agrees that they are curious animals. 'If you mimic a joey calling, you can often get the girls to come down the tree. And if you bellow, the boys will often come and have a closer look,' she tells me.

One of the local rangers who works in the park near my home thinks so too. 'We do surveys across the parks, cutting through bushland and not sticking to tracks. And it just feels like you're much more likely to see koalas sitting in the trees overlooking the tracks than you are in the more remote trees,' she comments.

It's not what you'd expect from a wild animal avoiding contact with humans.

'I think they get a bit bored and they like watching what's going on,' the ranger adds.

This is not to say that koalas are some kind of intellectual genius. But there might be a bit more to them than we usually give them credit for. Kath has worked with many marsupials – including some that are regarded as fairly intelligent, like wombats and the striped possums from northern Queensland.

'Koalas can habituate to routines,' Kath notes. 'They'll learn when you're coming to the door and can recognise individuals. But you can't train them to do anything. They just do what they want.'

I spent my student days training rats to press bars in Skinner boxes. If you can train a rat (or even a flatworm), I'm sure you can train a koala. It's all about patience, persistence and the right schedule of reinforcement. If you don't have the right treat, training is going to be difficult. Such koala 'sweets' – the leaves of *Eucalyptus utilis* and *E. platypus* – are regularly used by zookeepers to reward and train their koalas for public appearances, although admittedly only the ones who are amenable to do so.

Training koalas to respond to different stimuli is not just a way of using them for show. It's also an important non-invasive way of finding out what they can see, hear, smell, taste and feel. You have to tell the optometrist what you can see through different lenses – no-one else really knows what goes on inside your head. If we want to find out what's going on inside a koala's head, we must find a way of asking them, and trained responses for a reward are one way of communicating and learning about what they can perceive.

We're never going to know the cognitive capacities of koalas if we don't use the right techniques to study them. Until then, all we have are anecdotes and the intriguing hints of possibilities.

13

Sensory Overload

I want to know how a koala experiences the world. What it sees, hears, smells, tastes and feels. It's easy to assume that other species perceive their surroundings the same way we do – dominated by colourful visual stimuli, and oblivious to high- and low-frequency sounds, chemical scent trails, UV light or electromagnetic currents. It is not easy to shed our own perceptual biases and step into the sensory world of other animals.

We presume that koalas rely on hearing and smell – they have big ears and a large nose, after all. But there's much more to it than that. What exactly are they listening for and at what frequencies – high, low, ultrasonic or subsonic? What is it that they are smelling – each other, their food, the trees they climb or potential predators? How exactly does their vision work in terms of colour range, acuity, depth perception and adjustments for the dark? How do they feel the world around them – through their fingerprinted hands and feet, through proprioception, through their skin, fur and whiskers? And how have these traits been adapted for the peculiarities of their arboreal life?

If you go out with a spotlight at night in the Australian bush, or even in the outer suburbs of major cities, it is not uncommon to see glowing orbs of green, gold or silver from possums, gliders and kangaroos, as well as hares, rabbits and foxes. Koala eyes reflect bright silver orbs if you are lucky enough to spot one awake and looking at you.

Nocturnal animals usually have large dark eyes and pupils to absorb all the light they can, helping them see in the dark, with a reflective layer that causes them to glow in a spotlight. Although koalas also have this reflective layer for night vision, they have small eyes for their size – and particularly for a nocturnal arboreal animal. It's probable that koalas, living in the grey-green-brown world of the eucalypt forests, have dichromatic vision, like most other mammals, and can't see much of the red and ultraviolet at the ends of the colour range. Some marsupials, like the quokka and the southern brown bandicoot, can see in ultraviolet. Primates are unusual in having redeveloped tetrachromatic colour (like birds and reptiles), presumably for identifying fruit ripening into bright shades of red, orange and yellow in the sea-green canopy of trees.

For the most part, koalas have dark brown eyes as round as buttons, positioned to look forwards together, in binocular or stereoscopic vision, just like owls, cats and many primates. Occasionally, you might meet a blue-eyed koala, and then you realise that their eyes are not uniformly round and dark after all, but in fact have a vertical pupil like cat's eyes, which expands in darkness and contracts during the day into a narrow slit. They don't have much in the way of eyelashes, compared to the luscious ones of kangaroos, deer and humans. In other words, these are not the typical eyes of most tree-dwelling, nocturnal herbivores.

Just because they have small eyes, though, doesn't mean vision isn't important. One study put bags on their koala subjects' heads and found that the animals would not either search for or climb a tree in their enclosure.

'The koala appears to require a certain minimum level of vision before climbing trees,' concluded the paper, with dry understatement.

Animals that don't depend on vision don't have eyes at all. Small eyes, large eyes, round eyes, angled eyes, pupil size and shape are all adaptations for particular types of vision. Unusual anatomical structures always signify something interesting in biology, so what do these relatively small, lash-less, stereoscopic eyes with vertical pupils really tell us about what koalas see?

Large eyes with large round pupils are best for maximising low light, seeing in the dark and focusing on details. Being able to narrow your pupils to a slit helps reduce the amount of light entering the eye and allows you to see with better resolution, like squinting in bright sunshine or when reading the newspaper if we can't find our glasses.

Prey animals like sheep, deer and horses live in a wide, flat world that extends along a horizontal plain. They need to keep an eye on everything around them, maintaining constant vigilance for attack by predators, day and night. With eyes on each side of their head, their pupils narrow to horizontal slits, giving them near panoramic vision for more than 300 degrees of the horizon all around them. As the old adage goes, 'Eyes in the front have to hunt. Eyes on the side run and hide'.

Eyelashes too seem to split along predator/prey lines. Eyelashes are particularly long and thick in the plain-dwelling prey species, like cows, sheep, horses, camels and alpacas. Kangaroos and giraffes even have double-layered eyelashes of notable luxuriance, whereas predators tend to have either shorter eyelashes or none at all. Predators share this lack of adornment with arboreal animals like possums, lemurs, tree kangaroos and koalas.

Eyelashes, it seems, protect the eye from the drying effects of wind. At an optimal length, they act as windbreaks, diverting airflow away from the eye and reducing how often we need to blink. And interestingly, the list of animals with pronounced eyelashes is similar to the list of land animals that sleep the least. Eyelashes seem to help keep your eyes moist, clean and open for as long as possible. Small wonder, then, that koalas – masters of a bit of shut-eye – have little use for them.

Koalas live in a vertical world, not a horizontal one. They don't need to scan the horizon and they don't need to see across long distances. Their few predators come from above and below, rarely from the side. Vertical pupils focus on close detail and, along with stereoscopic vision, greatly enhance depth perception. The koala's eyes suggest that they need to be active both at night and during the day, and that what is most important to them is up, down and right in front of them.

I've again knocked over the koala skull that perches on my desk, and it has dislodged from the lower jaw. As I fit it back together, aligning the teeth so that they interlock into place, I have to navigate large bulbous structures that sit just behind the jaw. These two egg-shaped bulbs are the tympanic bullae, which house the middle and inner ear. These bulbs and their associated cavities are relatively big in koalas. In many ways they match the size of the koalas' ears, which, even without their fluffy fringe of fur, are really quite substantial.

Large tympanic bulbs are common in desert species – in both rodents and marsupials. Some desert rodents have ear cavities that are bigger than their brains. Such large ear structures also seem to be associated with detecting low-frequency sounds, which travel better over long distances, such as over open spaces in arid country, than in densely vegetated, wetter areas. The long reach of low-frequency soundwaves means they are also favoured by widely dispersed animals, like elephants and whales.

It seems likely that the koala's large tympanic bulbs and ears enhance their ability to pick up low-frequency sounds, like the bellows they use to communicate across long distances. But how does this work in the forest?

Forests generally muffle sound. Lush, damp and dense, with every surface smothered in soft ferns, mosses, mulch and moist, furry leaves, most forests absorb and deaden soundwaves. Noises suddenly disappear the minute you step inside a forest. High-frequency sounds attenuate first. Although low-frequency ones also become muffled, they last longer and travel further.

Australia's dry temperate eucalypt forests, however, are not soft. They are full of hard surfaces, driven by drought and low nutrients. These forests are a world of drought-hardened timber, leathery leaves and blistered bark: ironbarks, bloodwoods, stringybarks, mountain ash, mallee and jarrah. I imagine that the sounds in these forests bounce, refract and echo, shattering and splintering. A domain of continuous sclerophyllous sibilant whispers, percussive cracks, creaks and groans. Disordered and disorienting.

I suspect that those large furry koala ears rotating, filtering and sifting the sounds of the forest, like the 'dead cat' microphones used by roving reporters to dampen the noise of the wind. They are perfectly adapted to picking up the long-distance bellows of companions from windswept treetops.

One chilly winter morning, a monarch butterfly fluttered into Lana and Holly's enclosure at Cleland. When it landed on Lana's nose, her keeper Ash quickly snapped some photos, then sent them to the local paper. I wonder what made the butterfly land there in the first place. And what did Lana think of this strange visitor as she gazed, slightly cross-eyed, at the insect?

It's hard to miss a koala's nose. It's large, dark, flat and cute as a button, as the expression goes, particularly when a koala stretches out to bump noses in an endearing sign of curiosity or affection.

You can identify koalas by their noses too. Sometimes it is easy to recognise particular koalas through a scar, a way of walking, a pattern on their pelt or the shape of their face, but the patterns of light and dark skin in their nostrils are like interior freckles – as individually distinctive as their fingerprints, but easier to see, especially as they are constantly looking down at these puzzling creatures gathered beneath them.

Despite its large external layer, though, the nasal cavity itself is relatively small and without the complex folding inside the nose that most mammals have. This reduced surface area probably lessens water loss through evaporation – but it also means they can't lose heat as readily either. Koalas lack the glands that modify the temperature and humidity of the air. Nonetheless, their ears and nose both have a considerable blood supply, suggesting that their external size is important for cooling.

Despite the simplicity of their nose, the organs used to process smell are substantial. The olfactory bulb extends directly from the front of the koala's brain above the nasal cavity, absorbing and processing airborne chemicals. In addition, koalas possess a pair of extra olfactory organs, the vomeronasal organ, situated just above the roof of the mouth. In other animals – like snakes and lizards, cats and dogs – this organ detects molecules emitted by predators or prey, as well as sex pheromones from potential mates. For example, cats often produce the grimacing 'flehmen' response in reaction to sex pheromones from another cat, sucking air into the top of their open mouths towards the vomeronasal organ. Koalas, like other marsupials, probably use this organ to detect pheromones in the urine or scent-marks of other koalas, especially to identify females who are in oestrus and willing to mate.

Many marsupials scent-mark – with urine and saliva, and by rubbing glands in their chin, chest or cloacal area. Brush-tailed possums, some of the carnivorous dasyurids, red kangaroos and sugar gliders have all been observed doing this, although it hasn't been studied much except in sugar gliders, for whom scent-marking plays an important role in social cohesion by identifying members of their group.

For koalas, though, scent-marking plays a somewhat different role. It is as much to keep individuals apart as bring them together. Before climbing a tree, a koala will sometimes leave a few drops of urine at the base, as well as carefully sniff the area. Recent studies have demonstrated that males pay more attention to the urine of other males and of females in oestrus than they do to that of females who are not in oestrus. This lack of interest is mutual. Females who are not in oestrus avoid male koala urine, unlike the oestrus females.

The scent glands in the chest may play a similar role. Male koalas have a large gland on their chest, which is often visible as a dark smear around a bald strip in their white fur. Males over four years old mark the trees, particularly unfamiliar ones, they climb by rubbing this patch on the trunk or the branches.

'You can smell them when you're walking through the bush,' one of the local rangers tells me. 'Just like you can smell there's been a fox in the area, you can smell when a koala's been through – it's quite distinctive.'

Koalas can tell the difference between familiar and unfamiliar males from their smell. Scent-marking almost certainly provides males with a way of advertising their presence to rivals and females. This offers potential competitors and mates the opportunity to assess the male at a distance, without having to come into close contact and risk either an unwanted battle or an unwanted mating attempt – both of which seem to share some unfortunate similarities in koala behaviour. Better to sniff first and avoid any unpleasantness.

Eucalyptus is meant to be unpleasant to eat and the smell certainly deters plenty of browsers. But humans don't seem to mind it too much. We rinse our mouth with eucalyptus mouthwash, breathe in eucalypt aerosols, suck on eucalypt-flavoured cough drops, wipe our skin and clean our houses with it.

I break off a tiny fragment of fresh new gum leaf and carefully chew it. The pungent aroma of eucalypt explodes in my mouth, peppery and sharp. The older leaves are harder, tougher to chew and much more bitter. I spit out the leaf and rinse my mouth before picking up my coffee to drink it. As I take a swig, I realise that the coffee is much more bitter than the gum leaf. Aside from their refreshing eucalypt scent and general hardness, the leaves seem no less palatable than some of the bitter radicchio or mustard greens that grow in my vegetable garden.

Maybe a primate isn't the best judge of what's bitter or toxic. Coffee, chocolate, tea and chillies – we've built entire cultures around eating things that shouldn't be eaten, consuming the toxic plant defences and then perversely relishing the resulting pain.

If anything, koalas' sense of taste is even more secretive than their sense of smell. There are no conspicuous external organs to assess, and I can find no studies of koala tastebuds. But why would I? Koalas eat eucalypts. Surely they all taste much the same?

Clearly not, though, or koalas would not be so incredibly pernick-ety about the species, the trees and the leaves they choose to eat or ignore. How is it that they know which leaves contain optimal levels of water and nutrients – and, importantly, the lowest levels of toxins – if not through taste?

The only studies of taste I can find is genetic research suggesting that marsupials are more sensitive to bitterness than other mammals. Koalas, especially, also have active umami and sweet taste receptors. This sensitivity to bitterness makes sense. Presumably it allows the koalas to detect toxin levels in gum leaves and the nutritional sweetness of young sugar-laden leaves. But it's not entirely clear why koalas need increased sensitivity to umami flavours. Umami is the fifth basic flavour after sweet, sour, salty and bitter. The deep, earthy salty flavour is characteristic of the glutamate found in fermented soy sauces. Glutamate is a multifunctional amino acid utilised in many aspects of plant and animal metabolism, but it also plays an important role in plant signalling – in particular, it helps plants to sense when they are being eaten so they can launch a defensive attack.

I don't know if gum leaves have overtones of soy sauce, or if koalas can use this flavour to shift away from leaves that are about to become unpalatable. Suddenly, gum leaves sound like a much more interesting diet than I'd ever imagined possible.

And then there is touch. The oldest and perhaps least-appreciated of all the basic senses, it is apparent even in the simplest of single-celled organisms. Touch is the sense we rely on when all the world goes dark, to find our place, our support and stability. Emotional and physical.

Touch must be important for an animal with fingerprints and whiskers. Koalas may not have many whiskers, but they have more than we do – humans are one of the few mammals to entirely lack long, super-sensitive mobile facial whiskers. Whiskers seem to be of most value to animals that live in dark, complex worlds, like rats and dormice, for whom whiskers help guide foot placement when climbing, edge-following, and gap-crossing, as well as for close-up detection. Surely these are all tasks a treetop-dwelling nocturnal koala would have good use of whiskers for? Or perhaps whiskers are not as useful for finding the best leaves in a cluttered tree canopy as a sensitive nose and hands.

Of all the basic senses, touch is the only one essential to our normal healthy development. Babies can adapt to being born blind, deaf or without smell or taste, but children deprived of touch do not thrive. Like primates, koala joeys cling to their mother all through their infancy – even more so if you include their time in the pouch. I remember Harry Harlow's heart-rending experiments in the 1960s with maternally deprived baby rhesus monkeys, demonstrating the importance of tactile security. I cannot imagine anyone depriving a koala joey of touch.

Pioneering wildlife biologist and early koala researcher Ellis Troughton documented the story of the Faulkners, who adopted a three-month-old koala only to find that it cried constantly when left alone. Eventually, they provided it with a cushion wrapped in a koala skin, on which it slept contentedly, and later they gave it a large toy teddy bear to sleep with. Today, it is standard practice in koala medicine and rescue to provide towel-wrapped logs for post-operative or trauma patients to cling to. Throughout adulthood they cling to their trees, literally feeling their way through life.

My mother-in-law used to call the big gum tree in the park opposite her house the 'pat tree'. 'Don't forget to hug the pat tree,' she'd tell her grandchildren, as they walked to the playground. And the children would obligingly stretch their arms around the vast girth of the tree, resting their cheeks against its smooth polished bark while leaning against it, eyes closed and smiling. Even the adults could not help but stroke the tree as we'd pass it, feeling the tiny crenulations vibrate beneath our hands.

I cannot imagine what a koala senses when it leans into a tree – the security of its claws digging into the bark, the warmth of the sun or the cool of the shade, the firm reassurance of a branch at their back. Can they feel the freshness of the leaves they select, or the solidity of the branch bearing their weight? I don't know. But I do know that koalas, like all of us, need something to hold, whether it's a tree or the warmth of another body.

V

EVERYTHING CHANGES

⊘ The stories spread up and down the east-coast trading routes of the Yuin Nation, from Bidhawal Country in the south to Dharawal Country in the north. The first reports came from where the coast turned sharp from the cold southern currents to the warm waters from the north. A floating island growing great white trees had appeared offshore, close enough to see creatures clambering through the branches. Koalas, perhaps, or maybe possums?

It was Marrai'gang season. The cooler rains had started and the quolls could be heard calling for their mates in the forest. The people had begun their treks down towards the coast, their thoughts turning to renewing their fur cloaks and rugs with possum and koala skins as they watched this strange visitation.

A dark cloud settled out to sea and ancestral spirits rose from the depths in the form of three great waterspouts, which churned the turbulent waters. The ancestors danced and spun angrily across the sea, their transparent bodies flashing in a cloud of dark mist, keeping the strange island away.

The island disappeared to the north, and the people sent word by smoke and by foot up the coast about the approach of this disquieting apparition.

It approached again at Kioloa, close enough to see the tree creatures take human form, although whether they were pale as spirits or dark as people was not possible to tell. It was not an island at all, but a huge canoe – as big as a war canoe. The people lit fires along the beaches, the Elders gathered, marshalling their wisdom and knowledge as the canoe drew ever closer. The sea lifted in a heavy swell, throwing the huge canoe this way and that to push it away, but still it came closer. Just as it seemed about to land, the winds blew down from the mountains and pushed the canoe out to sea, towards the north again.

The canoe hovered, in and out of view, on the edge of the sea. The people kept watch along the shore, tracking its approach, and lighting fires at night to keep it at bay. The canoe taunted them, coming near then heading out again, then turning around and charging straight at them, like the angry whale Wondangar when the other animals stole his canoe. It came closer inshore at Woolungah, and a smaller canoe headed towards the beach, just like the one Kurrilwa, the strong-armed koala, used to escape from Wondangar and bring the people to land safely.

This canoe, though, was not filled with ancestors. Pale spirit creatures stood in it, calling and shouting, dead skin hanging loose from their bodies. The women hurried the children to the safety of the forests, while the warriors took up their positions behind the trees on the shore, ready to attack these demons from the other world as they came up the beach.

But the demons did not land. The sea gnashed its teeth on the shore, and the demons paddled away.

The creatures from the giant canoe finally landed a few days later at Kamay, on Dharawal Country. The pale ghosts did not seek permission to come onto Country. They did not sit on the shore edge and wait for the Elders to approach them. They stomped up the beach, making loud noises, ignoring the Dharawal men who told them to leave. They did not understand language. They threw pebbles and rocks, shouting all the time. They trampled through campsites, stealing spears and rifling through the huts and canoes. They chased the women and children watching from the trees. They left little bits of themselves behind on the ground, which hardened into beads and sticks that could not be broken.

There was no escape from them. They followed the Dharawal everywhere, to fishing sites and campsites around the bay. The ghost people dug little holes in the sand and chased everyone they saw. They hunted animals they had no right to kill and collected plants that could not be eaten. They spent hours making things that had no purpose. Some kind of sorcery. The Dharawal tried to ignore them. They covered themselves in white ceremonial marks, fashioned themselves protections, paid these malevolent creatures no mind in the hope that they would leave. The men threw their spears to send them on their way, but the demons shot stones with their fiery eyes, wounding some of the men. So, everyone kept their distance, and waited.

Finally, after eight days, their patience was rewarded. The demons retreated to their canoe and sailed away.

The ceremonies to cleanse Country took weeks.

Smoke from the ceremonial fires drifted up into the gum trees, where the koalas sat unperturbed, watching this unexpected activity. If they had any sense of foreboding of the tragedies about to unfold for the people below, for themselves, for the very forests they all relied upon, they kept their wise counsel to themselves.

They had survived worse calamities before. ᴐ⊸

14

Koalas Far and Wide

The river that runs through Adelaide was named by the Kaurna people for the river red gum forest that once followed its meandering course, from the hills across the Adelaide Plains to the wetlands on the coast. Karrawirraparri was said to reflect the path of the Milky Way across the sky. For millennia, the local inhabitants would have mapped that starry path across the plains with a mirrored trail of glowing campfires along the creeklines.

Koalas, too, must once have occupied these river red gum forests, as part of an interconnected population that spread across the south from the far west coast of Australia to the east. Although whether they shared these lands with the earliest people is uncertain.

What is certain, however, is that koalas are reclaiming these ancestral territories. They are following those same paths along Karrawirraparri and its creeklines, tracing the remnants of the river red gum forests back across the plains, into a world transformed beyond recognition.

Today, the starry path across the Adelaide Plains has dimmed and darkened. At night, the creekline wanders like a lost shadow through a sea of city streetlights. The red gum forest has been chopped down and reduced to a handful of remnant trees in intermittent patches along the banks. And the chain of seasonal billabongs – linked with gravel creek beds that once surged with winter floods and dried to a trickle in the hot summers – has been controlled, constrained and barricaded into a series of stagnant dams, drains and conduits known, somewhat euphemistically, as the River Torrens.

After two centuries of abuse and disfigurement, the river is slowly being restored. Parklands now stretch along its length. Introduced species have been reclassified as weeds and are being replaced with native plants. Swales have replaced drains, frogs fill reed-lined water-holes with harmonious vibrations, the river red gums and other eucalypts stretch their vast shady canopies overhead as joggers, walkers and bike riders share the paths beneath. And where the trees spread their canopies, the koalas have followed.

A friend of mine is passionate about koalas. We joke that she is the koala queen. She sees them on her daily walk through the linear park along the river, posting regular pictures of them on social media. I join her for one of her daily walks, to see her koalas. Allayne is slightly anxious that they won't show up.

'Some days there are heaps,' she says, 'and other days I don't see any at all.'

We wind through the park along neatly paved paths and well-maintained native garden beds and mown lawns. A handful of remnant trees are all that remain of a time long before busy roads hemmed in this creekline, before the plains were dissected, fenced, stripped, sealed and enveloped.

'There's one!' Allayne points. 'And look, there's a little one on her back.'

The koala watches us carefully as we circle beneath. Far from being oblivious to our presence, she is tracking our every move. She blocks our efforts to get a better photo, moving herself and her joey behind the tree branch and out of our line of sight. We decide to leave her in peace and head back to the path.

The koalas obviously like it here. And yet this is hardly the kind of prime habitat you'd think a sizeable wild animal would need to survive and breed. Despite the reedy billabongs chorusing with frog calls and the honeyeaters darting between flowering shrubs, this is not natural bushland. It's not a park for plants and animals. It remains, predominantly, a recreational space for people, not a place for koalas.

But here they here. And in ever-increasing numbers.

There are some droppings on the path – small, brown, oval-shaped cylinders packed tight with roughly chopped leaf fragments. I pick one up and crush it between my fingers to see what it smells like.

Allayne looks at me, puzzled.

'That's weird,' she says, laughing.

'Sorry – it's a biologist thing,' I say, trying to explain. Smell is often a key identifier for different mammal species. Koala scats smell strongly of eucalyptus, rather than the more pungent musk of many mammals. But these ones don't smell of much at all. They must be old. Fresh scats are shiny, look moist and have a strong aroma. They age rapidly, providing evidence for how recently an animal was there. Hunting for scats is one of the most common ways of surveying and locating an animal. Nowadays, they can even reveal the presence of stress hormones and enzymes, the koala's DNA, the trees they eat and the species of microbes in their guts.

Koalas are notoriously difficult to find unless you know what to look for. Scats on the ground are a key sign, but so too are the peculiar 'double' or train-track scratches that koala claws leave on trees. But koala-spotting techniques have become increasingly high-tech since my days in the field. Researchers now use specially trained sniffer dogs to locate koalas, and some are even trained to locate fresh scats. Thermal imaging from drones and planes has proven an effective, if expensive, method for locating animals in need of assistance after a bushfire.

I drop the scats to the ground. Allayne offers me some hand sanitiser.

'Whatever floats your boat, hon,' she says enthusiastically. 'Come on, let's find some more.'

We see several more koalas on our walk, and Allayne is delighted we have spotted so many. We head back to the carpark and into the torrent of traffic that sweeps along bitumen arteries, through a world full of straight lines – brick walls, concrete kerbs, steel and glass, power poles and wires. This is not the world that koalas have adapted for. And yet they, like some other wild species, seem determined to make themselves a home here, despite our inhospitality.

It's peak hour by the time I get to the main road, now crammed with cars. I glance at the GPS as I wait for the traffic to move, and I notice that the park behind me is one of the few curved lines on the map, a corridor of green snaking across the city from the forested hills all the way across the plains to the sea. This is the path the koalas are following – recolonising a country they have not lived in for thousands of years.

Not so much an invasion as a homecoming.

The story of the last three *Phascolarctos* koalas could be an Australian variant of the Norwegian folktale about the three billy goats gruff who, in search of new pastures, cross a bridge controlled by a fearsome troll. In the original story, the smallest goat and then the middle-sized goat persuade the troll not to eat them, but to wait for their larger, tastier brother – who instead defeats the troll, allowing all the goats to cross the bridge safely. The koala version of the story is also set in a near mythological world of long-vanished megafaunal species – reptilian 'dragons' and lumbering marsupial giants – but in this version, only the smallest member of the trio survives.

Two and a half million years ago, three koala species – one large, one medium and one small – also faced dwindling food supplies. Australian forests were retreating and retracting due to cold, dry and windy glacial conditions punctuated by milder, wetter and warmer inter-glacial periods. As the Antarctic glaciers froze, sea levels fell, exposing the Australian continental shelves and connecting the mainland to Tasmania, in the south, and New Guinea, in the north. Across the continent, these brutal cold droughts began replacing the forests with deserts and dunes. The southern forests survived on the tip of Western Australia but disappeared along the coast of the Great Australian Bight, forming the vast, virtually treeless expanse of the Nullarbor Plain, which now separates the western forests from the east.

The largest koala, *Phascolarctos yorkensis*, disappeared first, unable to adapt to their changing food supplies. As Australia swung back and forth through the Pleistocene climate oscillations, the eucalypt forests continued to wax and wane and diversify, adapting to the increasing aridity. Through the early and middle Pleistocene, only *Phascolarctos stirtoni* and *Phascolarctos cinereus* remained in the forests, their fossils being found in deposits – sometimes in the same time and place – in southern Western Australia, South Australia, Victoria and Queensland. By the late Pleistocene, the middle-sized koala, *Phascolarctos stirtoni*, is recorded only twice: first, in the Lake Eyre region of South Australia around 110,000–130,000 years ago, with the last record of its existence being 53,000 years ago at Cement Mills in south-east Queensland.

From that time onwards, the smallest modern koala became the sole heir to the remaining eucalypt estate. A significant part of this success must have been due to the modern koala's microbial digestive collaborators – a unique and flexible army, capable of adapting to suit the increasing diversity of eucalypt species, each with unique nutritional advantages and toxic burden. The koala itself no longer needed to accommodate the dynamic nature of the forest; instead, its built-in bacterial ecosystem adapted for it, like a personalised skeleton key able to unlock the nutrients of a wider and shifting array of food tree species.

The late Pleistocene, however, heralded another significant change to the Australian landscape. As the climate dried, the forest fauna retreated to live around more stable water sources: the rivers, creeks and ancient paleolakes dotted across southern Australia. And along these same pathways came the first people, the ancestors of Indigenous Australians, spreading out rapidly across the continent. This potent combination of retracting forests and a shrinking supply of freshwater coincided almost precisely with the arrival of this new competitor and predator, compounding the pressure on struggling megafaunal species in some areas. For the most part, though, climate and resources – food and water – dictated the fates of the megafauna, the koalas and the humans – all of whom found themselves in much the same dire predicament.

Even as the larger megafaunal creatures of the Pleistocene vanished, the smallest remaining koala survived. Although not without difficulty. Between 30,000 and 40,000 years ago, koalas underwent a significant population bottleneck – almost at the height of the last glacial maximum when conditions were at their coldest and most arid. So dramatic was this population decline that it has left its fingerprint in the genome of the modern koala, lowering its genetic diversity along with its abundance. Slowly but steadily, the last remaining populations of koalas in the west disappeared.

The last koala fossils found in the Margaret River caves of southwest Western Australia date from 25,500–43,000 thousand years ago. Fossil koala remains in the Madura Cave of the Nullarbor Plain are at least 24,000 years old. And koala remains accumulated in the Seton Rock Shelter on Kangaroo Island in South Australia – possibly as a result of scavenging by the dasyurid predator, the Tasmanian devil (now itself extinct on the mainland) – are probably between 16,000–21,000 years old. Kangaroo Island, it seems, may well have been a last refuge for several megafaunal species.

There is no evidence, fossil or otherwise, that koalas have ever lived in the Northern Territory, the northern regions of Western Australia nor in Tasmania. Despite the land bridge to Tasmania, Bass Strait remained a barrier for koalas – perhaps not being suitably forested to support their dispersal. For now, the koala is regarded exclusively as a species of the south and east coast forests, running in a coastal strip down the eastern seaboard from as far north as Cooktown in Queensland, across eastern New South Wales, throughout Victoria and into the south-western corner of South Australia.

Although how long that will last is anyone's guess.

The state borders have slammed shut again. Another COVID-19 case has escaped the haphazard hotel quarantine system. We are in hard lockdown for a few days, although with any luck the local cases will be isolated, any spread quickly identified, contacts quarantined and the rest of us will be able to resume normal activity.

But not interstate travel. Such is life in a pandemic. The COVID-19 coronavirus has swept across the world with startling, although not unexpected, rapidity, killing millions, overwhelming health services and incapacitating economies.

For a while Australia was fortunate, on an island continent with a relatively low-density population, moderately well-distributed wealth and health care and a high level of social compliance. Our borders have long been regulated for agricultural pests and diseases: a system readily extended, at times, to selectively excluding people from particular countries or cultural backgrounds.

As a nation of tightly federated states, our internal border controls are less strict. The flow of fruit, vegetables and plant matter is restricted from east to west to protect the more isolated and often disease-free industries of the south-west. But now it is people themselves who carry the disease, and whose movement must be limited.

I had planned research trips to Queensland, New South Wales and Victoria, to meet with various experts on local koala populations and go on field trips to better understand the variation in those koala populations, how they live and what they look like. In my youth I lived in all the eastern states and spent much time in their diverse forests, but I don't remember seeing koalas in the wild when I lived in Queensland or New South Wales. If I did, it was probably just a passing observation of a distant furry bottom, high in a tree. Not much to go on. I am most familiar with the southern populations – in Victoria and South Australia. But these are very different beasts, with very different histories from their northern counterparts.

In the past, people claimed that there were three subspecies of *Phascolarctos cinereus*. The smallest subspecies, *Phascolarctos cinereus adjustus*, was found throughout Queensland. Queensland koalas rarely weigh more than 6.5 kilograms and have a thinner, lighter pelt and a slightly different muzzle shape. *Phascolarctos cinereus cinereus* was distributed through New South Wales. The Victorian *Phascolarctos cinereus victor* was the largest subspecies, with males weighing up to 15 kilograms, with a thick, warm pelt. Curiously, the divisions between the subspecies fell along the administrative – and biologically implausible – lines of state boundaries.

A friend who is a wildlife vet travelled to the rugged and spectacular Pilliga region in northern New South Wales, where koalas, once feared wiped out by drought, seem to have recovered their numbers.

'They look completely different to the ones you see in Victoria,' she says. 'They are so small with little pug faces.'

It's possible that this variation reflects the latitudinal gradient found in many other species – known as Bergmann's Rule. Mammals of the same species tend to be larger the colder the climate is and the closer they are to the poles, as bigger animals are better able to conserve body heat than smaller ones. Not only that, but they also have thicker coats and sometimes have shorter limbs. Red foxes that are close to the equator, for example, are thinner and smaller than those in higher latitudes nearer the poles.

While higher temperatures do seem to greatly affect fur length and body size in Australia's northern koala populations, it's not the only factor. Body size can also be strongly influenced by diet. Perhaps the southern forests, with their larger stands of preferred eucalypt species, provide better-quality food for koalas, resulting in a higher density of bigger animals. More koalas in a smaller area increases fighting between males for mates, with bigger males having a competitive advantage.

I am frustrated by not being able to see these population differences for myself.

Like the koalas now trapped in isolated pockets of habitat, I have had my freedom to travel – wherever I want, whenever I want – suddenly taken away. It's a privilege I took for granted.

I guess if I cannot see living eastern koalas in person, I will have to confine myself to looking at their pelts. I email the collection manager for mammals at the South Australian Museum and make an appointment to visit the collections.

'So, we have a wide range of skulls,' says David Stemmer, as we head down a long, narrow corridor beneath the rabbit warren of buildings behind the museum's grand entrance. 'Which ones do you want to look at first?'

'Actually, I'm particularly interested in the pelts – the study skins,' I reply. 'I know you have some Queensland ones. Is it possible to look at them?'

'Oh, you want to see skins? That's great,' says David, looking pleased. 'We don't often get requests to look at them.'

He opens a large cupboard. An array of furry pelts hangs from coathangers on one side. Stacked on the open shelves opposite is a pile of furry, vaguely koala-shaped cylinders with small beige tags identifying their location, donor and date in varying degrees of detail.

These are not specimens for public display. They are for research purposes: a collection that, combined with those of other museums, represents the diversity in any given species. They are snapshots of changes over time and space. Most of them are from the twentieth century and, having been stored in darkness for much of that time, few are likely to have faded with age. While the oldest might have been hunted, most will have died either from natural causes, in the wild or in captivity, or perhaps from a dog attack or a car accident.

'You'll have to check the tags to find the Queensland ones,' David says, handing me a pair of gloves.

I run my gloved hand across the pelts. They are in assorted shades of grey and brown: some small, some large, some thick and soft, others short and coarse. I check the tags but none of these differences seems related to location or gender.

I notice a small grey koala with soft fine fur – a male – collected on 15 April 1918 by Miss I.M. Waddington somewhere in Queensland.

'Do you know who Miss Waddington was?' I ask. David shrugs and shakes his head.

This collection is too small for me to see the latitudinal gradient that has been found in larger studies. In any case, whatever the cause of the variation in koalas across the country, modern molecular studies have confirmed that all the koalas in Australia belong to the same species. While there are four genetic lineages apparent in koala populations, they do not conform to the historical subspecies defined by state borders. In fact, there are two lineages north of Brisbane, one central lineage on the New South Wales–Queensland border and one large southern lineage encompassing most of New South Wales, Victoria and South Australia.

In the past, it seems, the Queensland koalas formed one continuously interbreeding population, with an occasional barrier across the Brisbane Valley in southern Queensland and the Clarence River Valley in northern New South Wales. Whether these barriers were caused by the rivers, mountains, aridity or the nature of the forests is difficult to say. In general, though, the patterns in genetic variation suggest that, for most of their evolutionary history, koalas have been able to maintain gene flow throughout their population – until recently. At a smaller scale and over the last 200 years, local populations of koalas have become increasingly isolated and fragmented. They are no longer able to interbreed with neighbouring populations. It has not been glaciers or deserts that have had an impact on koala genetics and their long-term viability – it has been colonisation and the dramatic changes it has wrought on the Australian landscape over the last two centuries.

Whatever the explanation, they are all the same species – whether they are the hefty, thick-furred koalas of the south, or the small, thin-coated koalas of the north – and their future relies upon managing them all as a single species.

Mapping the distribution of koalas is not as simple as you might think. It's not easy to spot them, particularly in thick forest or in widely dispersed populations. And maps reflect a wide variety of purposes: historical and prehistorical distributions, where koalas are currently found or sometimes where people think they should be found. They differ by state and by region, depending on whether koalas are deemed to be endangered in one place, or introduced and overabundant in another. And their accuracy depends on the methodology employed and how many environmental managers and researchers, koala advocates or citizen scientists are recording their koala observations.

Some maps show koala distribution in a broad belt down the east coast of the country. Others are reduced to a scatter of isolated pockets with a larger splodge here and there. I cannot tell which ones are more accurate, but I do know that few of them map the South Australian distribution of koalas accurately. Perhaps there are better maps tucked away in government departments but if there are, they are not readily available.

The maps don't tell me which forests koalas are found in, or which particular trees within that forest, or which specific soil types or rainfall patterns they are associated with, or their distribution around free-standing water or rivers or aquifers – and how that differs in different areas. And few of the published maps differentiate between an area with one koala living among a hundred trees and an area with six koalas all sitting in the same tree. And yet surely that matters?

Some koala populations are studied thoroughly – others virtually ignored. How will we know if they are declining, or where they are disappearing from, or how best to protect them, if no-one knows where they are or how many there were?

Sometimes it's not even clear if the koalas we have are meant to be here.

'So, are koalas native to Kangaroo Island or not?' a friend asks me.

'Sort of,' I reply. 'It depends on your timeframe.'

It's a complicated answer. They are not native if you only consider the last few hundred years, but in the lifetime of the species they certainly are. We know they were present on Kangaroo Island at least 20,000 years ago, and they might have been around much more recently than that. Species do not go extinct overnight. They disappear gradually, surviving in small enclaves here and there, their ranges slowly shrinking, their breeding success declining, the numbers reducing until finally only a few pockets linger in isolated remnants of their former range. The last refugia of the south-west coast koalas might have been Kangaroo Island, or it might have been somewhere in the Mount Lofty Ranges, or the tips of the two western peninsulas, or even pockets of the diverse Western Australian forests themselves. We can't be sure. Fossils are not abundant enough to tell us for certain.

Like the thylacines that are constantly reported to still exist in the Tasmanian forests, could koalas have lived in the west in recent history?

In 1840, a young explorer by the name of Edward Eyre was tasked with finding an overland stock route between Adelaide and Albany at King George Sound in Western Australia, providing a cheaper alternative to coastal shipping. Eyre eventually completed his arduous journey across the hot, dry, treeless plains of the Nullarbor, around the Great Australian Bight, with the assistance of Wylie, an eighteen-year-old Menang man from King George Sound. At the head of the Bight, not far from Fowlers Bay, Eyre recorded remarkable evidence from the local Wirangu people that koalas still survived in the forests to the west.

> With respect to hills or timber, they said, that neither existed inland, but that further along the coast to the westward, we should find trees of a larger growth, and among the branches of which lived a large animal, which by their description, I readily recognised as being the Sloth of New South Wales; an animal whose habits exactly agreed with their description, and which

I knew to be an inhabitant of a barren country, where the scrub was of a larger growth than ordinary. One of the natives had a belt round his waist, made of the fur of the animal they described, and on inspecting it, the colour and length of the hair bore out my previous impression.

This account notes current knowledge, supported by a prized pelt that must have travelled along southern trade routes. Koala fur is very distinctive. Does Eyre's story suggest that koalas were present in the forests of the west within the last few hundred years, in addition to the tens of thousands of years ago indicated by the fossil record?

If anyone would know, it would be Indigenous people of the south-west area. Indigenous cultural memory is extraordinarily long and accurate – documenting, with incredible precision and detail, events that are known to have happened up to 10,000 years ago. It is one of the most conserved and robust systems of preserving knowledge in the world. Did koalas survive long enough in the west to feature in ancient lore or Dreaming stories? All of the First Nations stories I know about koalas seem to be from the eastern states. I ask some Western Australian contacts if koalas feature in Noongar oral history. They don't come up on any lists for the different language groups.

'I'd suggest Eyre was wrong,' replies Kim Scott, whose novels explore Noongar language, history and the impacts of colonisation. 'Those belts are made of possum fur, as I understand. Talking across languages, possum and koala would likely be similar.'

The Western Australian biologists who have worked with Indigenous communities to document their knowledge of mammal populations agree. They are adamant that Eyre must have misunderstood what the Wirangu people were telling him. Eyre may well have encountered, or heard about, koalas during his brief time farming sheep near the Hunter River in New South Wales. He listed 'sloths' among the chief and abundant edible riches of the country – perhaps he just assumed they would also feature in the Western Australian forests, not realising how different this isolated little pocket of south-west biodiversity is from the rest of the country. If koalas once numbered among Western Australia's riches, it has clearly been a long time since they vanished.

Nor, it seems, are they remembered in South Australia. In the 1970s, South Australian Museum staff took specimens and skins of South Australian mammals north, into the Indigenous communities of the Flinders Ranges. In a collaborative research project, they compiled the local knowledge for each of these species – which ones were still current, which ones had vanished and when. But there were no words for koala, no kindling of cultural memory, no ancient stories dating back generations.

All memories of koalas seem to have been lost here, along with the vast forests that once covered the hills and upon which these creatures so exclusively depend.

15

A New Arrival

It's summer, and clear blue skies shimmer with heat rising through the eucalypts. The grass has dried golden under the blistering sun; leaves crack beneath each footstep. Birds sit quietly in whatever shade they can find, and even the insects hum with a slow lethargy. It's increasingly common for Australian summer days to exceed 40 degrees Celsius for over a week straight.

The koalas do not like the heat. They sprawl belly-down over branches, limbs dangling apathetically below. Others slump in forks, legs outstretched in front and arms dropped to their sides. It is a wonder anything so relaxed can possibly maintain their balance. Some have slid to the ground, collapsed against the cool smooth trunks, or have taken up residence in trees offering thicker, denser canopies with better shade than eucalypts. Queensland koalas have thinner pelts to help them cope with the constant tropical heat, but the thick fur that southern koalas need to keep warm in winter isn't shed in summer and prevents them from cooling down. Thermal regulation is a constant struggle.

The zoos rig up shade cloths and spray misters to keep their residents cool. The green of well-watered lawns often attracts wild koalas out of the forest and into gardens and parklands. Home-owners leave buckets of water out for koalas to drink from, tip over or even climb into.

The name 'koala' is often said to be an Indigenous word meaning 'no drink', although from which of the more than 250 Australian Indigenous languages is rarely specified. Like all good stories, it's not entirely accurate either.

Koalas go by as many different Indigenous names as there are languages through the east coast forests, but those from around Sydney were called gula, or gulawarn in the Dharuk language, variously transcribed as cullawine, coolabun, koolewong, colo, colah, karbor, and koolah, all of which were eventually miswritten with an extra 'a' to form koala.

It's true that koalas have adapted to cope with very little water, and that much of their water requirements come from moisture in the leaves, but they are often observed drinking from freestanding water and even sitting in waterholes during summer and in heatwaves. In damp forests, koalas cluster along the trees on the rivers and creeklines. In arid habitats, access to freestanding water becomes even more important. Water makes everything better for koalas. It improves the trees they feed upon, making them less toxic and more nutritious, and it reduces the distance koalas need to travel. The energetic balance of their life works better with water.

Water brings life in the dry areas of Australia, regulating the breeding patterns of many birds and mammals – they breed quickly in times of plenty, and less when resources are scarce. It's a useful strategy in an unpredictable climate like Australia's. Being closer to freestanding water, like dams, has decoupled many species from this constraint and caused their population to boom, such as the large grazing kangaroos. I wonder how much impact this increased access to water has on koalas, and whether this might at times explain their increasing abundance on the edges of our major cities.

Koala may or may not mean 'no drink' but water has played a crucial part of koala evolution and ecology throughout their history, a fact which was not lost on the first people to come into contact with them.

The Dreaming stories of Indigenous Australians are complex embodiments of law, knowledge and culture. They reflect a long history in a challenging environment and contain sophisticated models of resource management and social regulation. They reveal just a fraction of Indigenous cultural knowledge, which is not shared or public but handed down, earnt and respected, from Elders to youth, through men and women, through lineages and totem groups and families. They are a small part of a system of oral history and cultural practice embedded in landscape and environment, intimately connecting people with the plants and animals they live with.

The value, effectiveness and even authenticity of oral storytelling tends to be underestimated in cultures that that have relatively recently started using writing as their major recording method. History is literally defined as 'written history' and dismisses any knowledge or events before or outside of written culture as a lesser kind of 'prehistory'. As readers, we see writing as some kind of true record, even though we know that the writer was a person, just as fallible and prone to error, misremembered details, bias and even deceit as any storyteller. We forget that for most of human history writing was a limited and closely guarded way of keeping secrets and controlling knowledge, accessible only to a tiny proportion of literate and wealthy individuals.

Oral cultures have their own ways of ensuring accurate replication – such as independent lines of transmission through separate clans, groups or genders, crosschecking through shared rituals with neighbours and the gradual accumulation of knowledge through ceremony and initiation over the course of each person's life.

Some stories are open and public – to be shared with the youngest child or the most uninformed stranger. Other stories are sacred and privileged, belonging only to certain individuals – the eldest and wisest members of a community. Outsiders may never be able to fathom the true meaning and value of the ancient stories that have been passed down through Indigenous communities across Australia, but even so, these stories provide a compelling reflection of a vastly deeper, longer and more sustainable engagement with the Australian environment than colonists, settlers and migrants have been able to achieve in a mere 200 years.

As a child, I loved reading Dreaming stories. My school library had a large, dramatically illustrated copy of *The Dreamtime Book,* almost too heavy for me to lift from the shelf. It was filled with creation stories collected from Indigenous Elders, particularly across central and northern Australia, and collated by the anthropologist Charles P. Mountford. These stories provided vivid accounts of how ancestral beings – sometimes human, sometimes animal, plant or geographic feature – created the people, places, plants and animals of Australia. There were stories of rainbow serpents, of the stars, sun and moon, of fire, droughts and floods and, of course, the animals that shared the bush where I lived – emus, willy-wagtails, cockatoos, kangaroos, sleepy lizards and abundant snakes. In my mind, these stories seemed more relevant than the demons, fairies, gods and angels that came from colder, wetter lands than ours. These stories belonged to the hot, dry country I had grown up in, with its eucalypt forests and fierce southern seas.

One of those stories was that of Koobor the drought-maker. Mountford didn't document who the stories belonged to or where they came from, but the story of how koalas acquired their ability to go without drinking water is told by the Wurundjeri people of Melbourne. In this account, an orphaned boy named Koobor was mistreated by his relatives. He was never given enough food or water. He learnt to live on gum leaves, but he was always thirsty.

One day, his relatives forgot to hide their water buckets when they went out foraging and, after drinking his fill, Koobor took all the buckets up into the tallest tree, far out of reach. When the group returned, they shouted angrily at Koobor to give them their buckets back but Koobor refused, forcing them to go thirsty too.

Eventually, some of the men managed to climb the tree and they beat Koobor, throwing him to the ground. As he died, Koobor changed into a koala and climbed back up the tree, where he stays to this day. He made it law that although koalas could be hunted and eaten for food, their skin could not be removed, nor any bones broken before the meat is cooked. If this law is disobeyed, the dead koala will cause a severe drought and everyone except the koalas will die of thirst.

As I grew older, I realised that these stories did not belong to me as much as I would have liked them to. I began to understand that – unlike the other stories in books that I devoured, absorbed, integrated and made my own – these Dreaming stories carried important cultural meanings for my Indigenous classmates that I needed to respect. I'm not sure I can ever lose that early fascination, but I can at least recognise the limitations of my knowledge and accept that I can only read these stories within the narrow framework of my own cultural background and expectations – in this case, from within the lens of ecological expertise.

The story of Koobor carries echoes of other stories I've heard about koalas. Not just their resilience in the face of drought, but also their ability to fall from great heights out of trees, to be struck on the head, thought dead and yet still survive, getting up and walking away. It references their incredibly tough skin, near impossible to remove from their bodies. The more carefully I read this story, the more I realise that along with its broader cultural value, this story is a compendium of environmental knowledge. It makes me wonder if perhaps broken koala bones have a tendency to make the flesh bitter when cooked, the way those of some other game animals do. Although that's not a question I wish to find the answer to.

There weren't many koala stories in Mountford's book, which focuses on inland desert country. But there are many stories about koalas from the east coast. The story of Wondangar the Whale and Kurrilwa the Koala narrates how the Tharawal (or Dharawal) people first arrived in the Lake Illawarra area, about 80 kilometres south of Sydney. It tells of how people had heard stories of this fertile land from migratory birds and wanted to settle there, but they lacked a canoe big enough to make the ocean crossing. So, with the help of the Starfish, they tricked Wondangar the Whale and paddled away in his giant canoe. Wondangar chased them when he realised their deception, and they only escaped through the strong arms of Kurrilwa the Koala, who was able to keep paddling to safety ahead of the angry whale.

The Gumbaynggirr people north of Sydney also tell a story about koalas helping people across the water. There are many versions of the Dunggiirr koala story, but most accounts involve a long bridge made of koala gut to rescue people cut off by rising sea levels. Elements of this story echo other ancient stories around Australia that describe the flooding of coastal plains, inlets or bays off the eastern seaboard over the past few thousand years.

Current Gumbaynggirr storyholder Aunty Shaa Smith recently shared this 'storyplace' in a research paper explaining the relevance of such stories today. Aunty Shaa describes how the mischief-making Baalijin, or eastern quoll, tried to sabotage the Dunggiirr koala brothers' efforts to save people, by threatening to cut down the bridge, stirring up the seas and calling Gurruuja, the great whale, to frighten them. This story describes a world out of balance: Baalijin's destructive energy disrupts the Dunggiirr brothers' efforts to use their creative clever energy to try to restore the balance.

If we fail to listen to the Koala Brothers, to heed the customary law of traditional owners, these forces remain out of balance and we fail to learn how to live in a changed world, in balance with the land and nature. It's a lesson perhaps even more pertinent today than it was in the past.

We have much to learn by listening carefully to what koalas can tell us.

Near Carnarvon Gorge in central Queensland, the local Bidyara people regard the koala, Didane, as a wise counsellor who turned their barren lands into a lush forest. According to the Bidyara story, the people tried to make the trees that were growing in the sky release their seeds so that they would fall to earth and cover the barren land with plants. No matter how hard they tried, though, they could not throw their boomerangs that far, and their boomerangs returned to earth without any seeds falling. They called on the strong-armed Didane to help them. He brought his biggest war boomerang and hurled it with a mighty effort into the sky, where it disappeared into the clouds. The boomerang did not return and the people feared that Didane had failed, but then seeds began to fall from the sky and coat the ground. And with the next rains, the forests flourished and the people rejoiced.

On my coffee table, I have a tray full of different 'gumnuts' – the hard, woody fruit of the eucalypt trees – that I collected from the Currency Creek Arboretum. They come in all sorts of shapes and sizes – huge bells, tiny nodules, large spiky sputniks and ones that look like winged rockets. As I pick up one of the nuts, a shower of powdery granules falls into my hand. I had forgotten that the gumnuts typically hold onto their seeds until they are completely dry – after extended drought or even a fire – sometimes waiting for years before releasing them. Perhaps they are waiting until circumstances are so dire that the death of the parent plant is imminent. But whatever the trigger, they will lie dormant in the dry soil until moisture triggers their germination and the renewal of the next generation in the eucalypt forest.

I can't help but think how apt it is that the koala – sitting in the tree, surviving drought and hardship – should be seen as the harbinger of this shower of seeds that brings revitalisation to the forests on which they depend.

The arrival of the first humans in Australia exposed the koalas to new predators and new impacts. Humans are a significant force in the landscape wherever they live, moulding and reshaping it to suit their own needs, using fire, axes or digging sticks to increase their food supply, opening up trails for trade, travel and migration, and modifying the survival and habits of the species they prey upon.

Humans are relative newcomers in evolutionary terms. The phascolarctids have been around for 24 million years; the hominids for just 6 million years. The oldest modern koala fossil is at least 350,000 years old, while the oldest *Homo sapien* fossil is no more than 210,000 years old. For most of their existence, koalas existed without any pressure from humans at all.

Even so, the impact of Indigenous Australians on the landscape was likely to have been slight – at least compared to what was to come. They lived at low densities (compared to current urban practices), dispersed over the country, occupying areas capable of supporting their smaller populations over longer time scales, through Australia's irregular and unpredictable climate, including lengthy droughts. While the use of fire may have had local impacts, the most dramatic changes in both wildfires before human arrival and subsequent patterns in Indigenous use of fires seem to have been caused by climate. These early human populations show no signs of boom-and-bust cycles – the evidence suggests stability, ecological balance and the evolution of long-term sustainability.

Indigenous Australians were proficient hunters of koalas, for food and for fur. But their use of this species varied not only by nation, clan and family, but also by the kinship system of moiety, totem and skin that defines relationships and responsibilities within Indigenous society. Europeans may have seen Indigenous Australians as all being the same and classified them as 'Australian Aboriginal' people, but with over 250 different languages and 800 dialects, the Nuenonne of Tasmania were as culturally distinct from the Menang of the southwest or the Wik people of the Gulf of Carpentaria as a Spanish person might be from someone in Turkey or Finland.

In the far south-eastern corner of Victoria – in the dense, dark Gippsland forests, where trees once grew over 100 metres high – the local Gunaikurnai men were impressively skilled tree-climbers.

'They used their tomahawks to cut notches in the bark,' observed John Bulmer, the manager of the Lake Tyers Mission Station in the Gippsland Lakes in the 1870s, 'which they use as toe holds to climb very high, straight trees. Sometimes they used a band made of stringy bark (yangoro) which encircled the tree and their bodies. This made it easier to cut the notches.'

Some of the group waited below, while one man climbed the tree.

'After much labour they got to the top of the tree, sometimes a very giddy height, but they were perfectly cool and proceeded to kill their game,' Bulmer reported. 'It often happened that the bear would get on to a very remote bough, when it had to be cut to let the animal fall, but there would be others below to dispatch it. After all their labour I have seen the bear leap onto another tree, where they must begin their work again or leave it. They generally did the latter.'

Despite sitting in trees, as if waiting to be picked off, koalas were perhaps neither as easy to catch nor as abundant as other more palatable prey sources. A Dharawal man from south of Sydney was observed in the 1830s hunting koalas with a long thin pole and a noose of ropy bark. As the man climbed the tree, the koala took refuge in the highest branches, too light to take a grown man's weight. The hunter used the long pole to try to hook the noose around the koala's neck, which the koala resisted by scraping the noose away with its paws. The hunter then climbed higher, the tree bending under his weight, until he finally managed to tighten the noose around the koala's neck and steadily pulled it down the tree with him, keeping a very careful distance from the irate koala's claws.

As soon as they reached the ground, however, the koala freed itself and leapt to the nearest tree, ready to climb to freedom again, but was knocked to the ground by a tomahawk flung with considerable force and accuracy by the Dharawal hunter.

It is sometimes claimed that hunting by Indigenous people kept koala numbers in check, or perhaps even dramatically suppressed them. This story originates largely from an account by Harry Speechly Parris (1885–1964), whose father had moved to Australia from England and occupied some 80 acres of land on the Goulburn River in Victoria in the late 1860s. In 1930, Parris published an account of local settlers' recollections of the dramatic changes observed in the area before and during his lifetime. He noted that in the 1870s the red gum forests surrounding the river had been densely populated with koalas. However, one of the early European settlers in the area, William Day, who arrived in 1854, claimed not to have seen a koala at all for the first three years of living there, although it seems that older residents had seen one or two elsewhere in the area.

'Inside a month,' Parris wrote of those first sightings, 'there were hundreds of bears in the locality, indicating a migration.'

Parris went on to report that prior to 1854, no koalas were present in the lower Goulburn, despite being abundant in nearby Gippsland, Upper Goulburn and the Dandenong Ranges. Initially, Parris did not believe that hunting by Indigenous people had had any impact on koalas. Rather, he suspected that disease might have caused the decline. But writing about this subject a few years later, Parris realised that some Indigenous people did hunt and eat koalas, leading him to speculate that this might have been the cause of their decline.

'The bears increased as the blacks decreased, and in the [eighteen] fifties and sixties they spread over a greatly increased area,' Parris claimed.

Parris was, like many Australians, concerned for the fate of the koalas, and he would not be the only one to link the fate of a species to the declining fortunes of Indigenous Australians. Although Parris wanted koalas to be reintroduced to the Upper Goulburn area and protected, his interpretation of local recollections would take on a life of its own and echo through the literature. From that day on, papers on koala management cited Parris as incontrovertible proof that hunting by Indigenous Australians had once dramatically supressed koala populations, which therefore supported the notion that 'plagues' of koalas required constant culling and hunting to maintain the balance of nature.

If ever evidence was needed of the dubious and inappropriate use of the written word as 'historical proof', poor Harry Parris's paper, exhorting us to take care of this 'best-loved of all wild animals', is it.

Just how abundant were koalas in different forests prior to European colonisation? Were their numbers curtailed by Indigenous hunting? Or did Indigenous hunting replace that by large, marsupial hunters like thylacoleos or thylacines? It's an important question that shapes the way we see, treat and manage koala populations. Are they a species that 'needs' to be controlled, or is their current boom-and-bust cycle simply an artefact of a resilient species trying to re-establish balance in a shattered and collapsing ecological landscape?

The haphazard reports of early European explorers and settlers are hopelessly skewed by their own preconceptions and cultural biases. They often did not see things that were right in front of them. And with the loss of so much cultural context, it is difficult to interpret the oral history of Indigenous Australians. Is there any other source of evidence for the prevalence of koalas prior to European colonisation and their importance in Indigenous hunting?

It occurs to me that Indigenous artefacts in the collections of our museums might provide some clues. If Indigenous Australians hunted koalas enough to affect their population, it could be reflected in the tools, ornaments and artwork they produced, using the animal's bone, teeth, fur and skin. I search the Indigenous collections of Museums Victoria for artefacts mentioning possums, koalas, wombats and kangaroos in their descriptions. There are sixty-two artefacts for kangaroos and fifty-five for possums, but hardly any for either wombats or koalas.

The only 'koala' item I find in the collections is a 'jimbirn' – a string headband made in about 1885. These headbands were once highly ornamented: strings of kangaroo teeth hung on possum fur at the temples, a wild dog's tail hung from the back, while eagle, emu and cockatoo feathers were added, along with the furry tips of koala ears. This particular one does not have any ornaments, not even koala fur. But koala ears were used as decorations among Gippsland communities, particular by young women.

By contrast, kangaroos and possums are abundantly displayed in museum collections. Indigenous people used kangaroo fur for bags and possum fur for warm cloaks. They transformed kangaroo tails into armbands, their teeth into necklaces, their bones into tools and their sinews into string. These animals were represented in paintings and carvings, on hunting tools and skins. There is clearly no evidence here that koalas (or wombats) were being hunted anywhere near as much as kangaroos or possums.

This pattern conforms with surveys of other cultural objects and artwork. A large study of privately traded carved artefacts from Indigenous people also found lower frequencies of koala images compared to other animals, around 5 per cent, rather than closer to 20 per cent for emus and kangaroos, although with regional variations. A colleague who works in archaeology tells me the same pattern is apparent in rock art.

'My understanding is that koalas are quite rare in figurative rock art. They are mostly noted in the Sydney Basin rock art,' Dr Daryl Wesley replies. There are petroglyphs, or ancient rock carvings, all around Sydney. Some may have been made at least 4000 years ago, but many have been continuously maintained and 're-grooved' since then and more were added until the arrival of European colonisers. Some are beneath houses and on streets, but many are in the bushland areas surrounding the city. No-one is entirely sure what these carvings signify, and the people who might have known were decimated by the smallpox the First Fleet brought to Australia. Even here, Daryl tells me, where rock carvings are common, images of koalas are rare.

One study of the Nepean River area, where koalas were historically abundant, found that they only accounted for two out of 3541 motifs. And across the broader Sydney Basin, which encompasses much of the south-central New South Wales coast, there was only one koala recorded in 7804 motifs. By contrast, kangaroos and wallabies were represented in 7 per cent of images and fish in 12 per cent.

'It's worth look at the archaeological sites,' suggests Daryl. 'If koalas were being hunted, their remains will be in the faunal assemblages. I can have a quick look through the papers for you.'

A day later, Daryl gets back to me.

'I've had a look,' he says. 'I'm quite surprised by the lack of koalas. Even where they are found in Queensland and New South Wales, the number of individuals is incredibly low, compared to kangaroos. Koalas certainly don't seem to be much of an important economic resource.'

Given their eucalypt diet, it's possible that koalas, while eaten from time to time, were not as popular as other prey items. It is possible that they were only eaten when other preferred prey like kangaroos and wallabies were not available. The value of their fur may also have been reduced by the difficulty of skinning them (a trait they share with the equally infrequent wombat). Across all the different measures – from the fossil record to art and artefacts – koalas represent only a small proportion of the animals found and a similar proportion to what we'd find among living animals in healthy sustainable koala populations today. It seems likely that koalas have always been a relatively rare, widely dispersed animal compared to the more common kangaroos and possums.

The ways in which Indigenous Australians used koalas for food or fur is every bit as diverse and complex as the many Indigenous nations that stretch across the country. There is no reason to assume that people in the coastal wet tropics of far north Queensland would share similar ecological practices to those in the fire-prone open woodlands inland or the dense, tall forests of the Victorian mountains. The only thing we can be sure of is that the hunting, eating and preparing of koalas – like all other plants and animals – was tightly regulated by a complex system of moiety, age and gender, which seems to have been designed to regulate the use and distribution of food resources to ensure sustainability in an often unpredictable and variable climate.

16

The English Annexation

The koala descends down the trunk carefully. The joey on her back fidgets, unbalancing her. It's half her size – a heavy burden to carry. Her ears flick back and forth, betraying her anxiety. As she reaches the ground, she sniffs and listens, then moves off before her offspring can think about exploring on its own. She sets out briskly, the joey bouncing heavily on her back. There is little undergrowth to obstruct her in this open woodland, but no cover either. The trees are generously spaced, their canopies separated and too far apart for her to move safely through the branches.

A whip cracks through the undergrowth, a distant sound of voices, of dogs barking. The koala picks up her pace, but she cannot travel much faster. She reaches one of the larger trees. It's probably not the one she was hoping for, but it will have to do. She clambers up the trunk, pausing between short leaps. Climbing is always an effort, even without carrying a heavy joey. She reaches one of the lowest branches and sits down to catch her breath. It's not high enough. As the horses, dogs and humans come closer, she moves further up, hopefully out of reach, out of sight.

The noisy party clamour beneath her tree. The dogs sniff at the base, where she stopped, yelping, circling and alerting their master.

She watches carefully through the leaves, but the man does not look up. He calls the dogs on. They are not hunting today. It's just on dusk and he wants to reach his destination, and his dinner, before nightfall.

One of the dogs gives a half-hearted howl before lolloping off after the horses. She would never have been able to outrun them, and her claws would have been of little use against so many teeth. They would have ripped the joey from her back and probably dispatched her too.

The koala waits in the tree until dark, her joey attempting to eat the unpalatable leaves nearby. She will wait until the moon sets and the forest is dark and still before moving on. Her dinner will have to wait.

In October 1785, a young Lancashire sailor by the name of John Wilson was convicted of stealing 'nine yards of cotton cloth called velveret, of the value of tenpence'. In 1788, as punishment, he was boarded onto the First Fleet and transported to the east coast of Australia for seven years.

Social upheaval was rife across Europe towards the end of the eighteenth century. Colder conditions during the Little Ice Age had made life increasingly hard for peasants and workers. Migration to North America and the Revolutionary War of Independence spread stories of new opportunities, wealth and freedom. In France, this social upheaval precipitated the French Revolution and the violent abolition of the monarchy and aristocracy. Faced with similar social inequity and the loss of their American colonies, vested interests in England directed their efforts to the Indo-Pacific, towards establishing a penal colony in Australia for the poor and political dissidents, and appropriating the vast wealth of China and India.

On one particular Saturday in 1788, eleven British ships officially began disembarking over 1000 men and about 200 women, as well as goats, pigs, sheep, horses, cows, rabbits, bed bugs and smallpox onto Australian soil. Aptly enough, they landed at what is now the Overseas Shipping Terminal at the heart of Sydney Cove, opposite the more scenic Sydney Opera House. The official landing of the First Fleet and the founding of a British penal colony on 26 January 1788 is today either celebrated as Australia Day or mourned as Invasion Day. But whatever we call it, it was indisputably an event that devastated the Australian landscape forever, beginning an avalanche of introduced diseases, predators, weeds, pests, mass extinctions, genocide, broadscale land clearance and deforestation, which has left a vast swathe of the surviving species and ecosystems threatened and endangered.

History does not record how Wilson found his seven years in the brutal penal settlement of Sydney, but it seems that once he had served his sentence, he craved more civilised company. He took to the bush and found companionship with Indigenous people of the surrounding country – the Eora and the Dharawal – living as they did for several years, apparently going by the name Bun-bo-e, wearing only a kangaroo skin and proudly displaying the scars of tribal markings. At the same time, he learnt much about the plants, animals and landscape of this unfamiliar country.

Wilson existed between two worlds – tolerated by both but belonging to neither – connected to the established Indigenous knowledge of Country but also to the invaders' camp perched on the edge of the continent. He became a conduit for information about the surrounding lands for the colonisers, and begrudgingly called upon to safely guide English explorers through the area.

Wilson wasn't the only one who thought that the Australian bush offered more promise than the confines of a penal colony. Irish political prisoners who had been sent to this isolated location by English authorities soon became enamoured of the notion that there was some kind of utopian settlement south-west of Sydney, somewhere near modern-day Bargo or Campbelltown. Wilson was tasked by Governor John Hunter with guiding four of the Irish convicts, under a heavily armed guard, to the area to find out whether the rumoured settlement existed.

Wilson left no record this expedition, but John Hunter's servant, nineteen-year-old John Price, did. On 26 January 1798, about 90 kilometres south-west of Sydney, near the headwaters of the Nepean River, Price recorded a singular discovery.

'The ground was very rockey and brushey, so that we could scarce pass,' Price wrote. 'We saw several sorts of dung of different animals, one of which Wilson called a Whom-batt, which is an animal about 20 inches high, with short legs and a thick body forwards with a large head, round ears, and very small eyes; is very fat, and has much the appearance of a badger. There is another animal which the natives call a Cullawine, which much resembles the Sloths in South America.'

It's not entirely clear from this whether Price himself saw a koala or just some dung, but it is clear that there are two sources of information in this account. Someone was well read enough to know about South American sloths. That was probably either John Price or Governor John Hunter, who had a considerable interest in natural history. But someone also knew what koalas and wombats were, what their scats looked like and what their local names were. This knowledge can only have come from the Indigenous people with whom John Wilson associated.

It had taken Europeans ten years since first settlement, and almost thirty years after James Cook first began exploring the east coast, to even notice the existence of unobtrusive koalas, a pattern of inattentional blindness that repeats through the historical records.

It is likely that koalas did not live in the area immediately surrounding the penal colony at Port Jackson because the soils and trees were probably not the right kind. But a survey map from 1791, published in Watkin Tench's account of the area explored around Sydney, clearly shows that plenty of journeys had been made across the Wianamatta shale soils to the headwaters of the Hawkesbury and Nepean rivers, where koalas and their favoured trees could be expected to be found in some abundance. Tench's map, though, is littered with descriptions like 'sandy, rocky and very bad country' or 'still wretched and brushy', 'miserable', 'dreadful', 'barren' and 'poor'.

Perhaps these colonists, seeking to re-create the farmlands and coal mines of home in this New South Wales, were more interested in what lay beneath their feet than what was sitting above their heads, watching them pass below.

Wilson's exploits in bringing the koala to European attention, like his early forays into the Blue Mountains, have largely been overlooked, no doubt because he was a convict who 'went native'. A more historically acceptable 'discoverer' soon came in the form of Francis Louis Barrallier, who, despite being French, was employed by Governor Philip King as an architect and later an explorer.

In early November 1802, one of Barrallier's forays took him south-west, across the Nepean River. Barrallier's expedition included four soldiers, five convicts and, importantly, a local man, Gogy, with

his wife and child. They travelled by bullock-drawn wagon with supplies for the depot at Nattai, some distance west of Campbelltown.

'After having travelled over a plain, I perceived fires in several
places,' Barrallier wrote in his journal on 8 November. 'Bungin told
me that it was a chief called Canambaigle with his tribe, who were
hunting, and had on that very day set the country on fire.'

When they arrived at the depot the next day, they met some
Dharawal people who had not encountered Europeans before.

'Gogy told me that they had brought portions of a monkey (in the
native language "colo"),' continued Barrallier in his journal, 'but they
had cut it to pieces, and the head, which I should have liked to secure,
had disappeared. I could only get two feet through an exchange which
Gogy made for two spears and one tomahawk. I sent these two feet to
the Governor in a bottle of spirits.'

He added a further footnote to 'colo': 'Gogy told me that this portion of the colo (or monkey) and several opossums had been their share
in the chase with Canambaigle, at which they were present on the day
when that part of the country which I referred to above had been burnt.'

Barrallier is also often associated with the first live koala to have
been brought back to the colony, but whether this is the case is
unclear. Barrallier had already left Australia in May 1803. Certainly,
his discoveries – and perhaps the strange appearance of the feet he
returned with – lead other collectors to travel south-west of Sydney
in search of more colo-monkeys.

On 21 August 1803, the recently founded local newspaper, *The
Sydney Gazette and New South Wales Advertiser*, recorded:

An Animal whose species was never before found in the Colony,
is in His Excellency's possession. When taken it had two Pups,

one of which died a few days since. This creature is somewhat larger than the Waumbut [wombat], and although it might at first appearance be thought much to resemble it, nevertheless differs from that animal. The fore and hind legs are about of an equal length, having five sharp talons at each of the extremities, with which it must have climbed the highest trees with much facility. The fur that covers it is soft and fine, and of a mixed grey colour; the ears are short and open; the graveness of the visage, which differs little in colour from the back, would seem to indicate a more than ordinary portion of animal sagacity; and the teeth resemble those of a rabbit. The surviving Pup generally clings to the back of the mother, or is caressed with a serenity that appears peculiarly characteristic; it has a false belly like the apposum [opossum], and its food consists solely of gum leaves, in the choice of which it is excessively nice.

These animals appear to have survived at least a month later, as they were reported in the same paper as being under the care of Sergeant Packer of Pitt's Row, who (ominously) is reported to have been supplementing their gum leaf diet with bread soaked in milk or water.

Despite enthusiasm for the strange Australian kangaroos in England, there seemed little interest in koalas. Robert Brown, the botanist on Matthew Flinders' voyage, encountered the captive koala while in Sydney, and his colleague Ferdinand Bauer began his exquisitely detailed and accurate drawings of the animals. But these were not published until the 1960s, and by and large English representations of the koala were based on a more rapidly completed painting by the artist John Lewin in 1803.

Several koalas were sent back to England, but none appear to have arrived alive, no doubt because of the lack of suitable food for the voyage.

The skins and skeletons of koalas, however, did make it back to England. These specimens ended up with the anatomist Everard Home, who published a few brief notes on the 'koala wombat' as part of his longer treatise on wombats in 1808.

For years, it was English gardeners, collectors and dealers, rather than scientists, who were enthusiastic about Australian natural history. Aside from Home, the only scientists interested in resolving the mysteries of marsupials seemed to be those in mainland Europe, but they lacked access to the specimens.

Once Everard Home had completed his anatomical studies of the koala, he appears to have passed the skins and skeletons on to William Bullock, the entrepreneurial owner of the 'London Museum and Pantherion' which eventually contained 'upwards of fifteen thousand natural and foreign curiosities, antiquities and productions of the fine arts, collected during seventeen years of arduous research and at an expense of thirty thousand pounds'. He displayed these animals in Piccadilly along with other curiosities from the little-known southern continent, including kangaroos and platypus, 'kangaroo rats' (potoroos), 'porcupine ant-eaters' (echidnas), 'spotted fitchets' (quolls) and the remarkably named 'zebra opposum' – a now extinct thylacine which Bullock states was found with an echidna partially digested in its stomach.

Bullock apparently had two koalas in his display, one larger and one smaller. While he devotes quite a lot of space to the wombat in his exhibition catalogue, the koala seems to have inspired less interest.

In the 1810–1811 catalogues, it is described in a single line: 'In this case is also a large animal from New Holland, called the Koala.'

Whatever this early representation of the koala looked like, it must have shaped European perceptions of the animal for decades to come.

Taxidermy is a highly skilled art. Preparing a lifelike specimen of a species which you have never seen alive, which is unlike any other animal you know and which you don't have any good drawings of must be a nearly impossible task. I have come across Australian animals in regional British museums before, tucked behind the ornate Pacific canoes and spears, and noticed a small, brown, wizened possum-like animal, crouched on a low branch, overshadowed by a large kangaroo. It had no tail and a face like a quokka, and its large once-fluffy ears were tattered and bald. It was, without question, the smallest, saddest, most implausible specimen of a koala I had ever seen. I'm not sure that the general British public gained any great insights into koalas from the display in Bullock's museum either.

Bullock did not include an image of the koala in his catalogue, but the naturalist George Perry almost certainly used Bullock's stuffed koala (as he did for several other species) as the model for his drawing in 1810, the first known illustration of the species to be published in Europe, in his publication *Arcana: The Museum of Natural History*. Perry's illustration depicts an animal about to climb a tree, arms fully extended and standing on its hind legs. Its ears are sparse and wispy; a flat black nose protrudes over an undershot chin, and glass-button eyes stare out at odd angles. The five claws on its hand are equidistant from each other, as are the four claws on its feet.

In his text, Perry described the characteristic double thumb and finger of the koala, even though they're not obvious in his picture. He noted that these digits were unusual for an animal he thought might be most closely related to a sloth or a bear. But he seems nonetheless persuaded by the strange misshapen appearance of the creature in Bullock's gallery.

Whether we consider the uncouth and remarkable form of its body, which is particularly awkward and unwieldy, or its strange physiognomy and manner of living, we are at a loss to imagine for what particular scale of usefulness or happiness such an animal could by the great Author of Nature possibly be destined ... The Koalo is supposed to live chiefly upon berries and fruits, and like all animals not carnivorous, to be of a quiet and peaceful disposition ... They have little either in their character or appearance to interest the Naturalist or Philosopher. As Nature however provides nothing in vain, we may suppose that even these torpid, senseless creatures are wisely intended to fill up one of the great links of the chain of animated nature, and to shew forth the extensive variety of the created beings which God has, in his wisdom, constructed.

With a determinedly Eurocentric focus, it seemed almost impossible for many European observers to believe that many of the distinctive animals of Australia might in fact be unique and entirely unfamiliar, with no allegiance of any kind to the creatures of the Old World. For the English, Australia was little more than a penal colony. Botany Bay and everything about it – its animals, environment, people and potential – would be shadowed by this peculiar legacy for many years to come.

Just as the first description of a eucalypt was by a visiting French scientist, who had examined the vast collections of Joseph Banks, the first scientific description of a koala was by a French scientist, Henri Marie Ducrotay de Blainville, who visited London in 1814 and provisionally named them *Phascolarctos* in his seminal paper on the classification of animals. After describing the koala's teeth, hands, feet and noting the lack of tail, Blainville continues: 'The size of a mediocre dog, this animal has long fur – bushy, coarse and chocolate brown. It has the bearing and gait of a little bear; it climbs trees with great ease. It is called Colak or Koala in the vicinity of the Vaupaum [Nepean] River in New Holland [Australia].'

Blainville hoped that a more complete description would appear in the forthcoming work on marsupials by his superior, Étienne Geoffroy Saint-Hilaire. Saint-Hilaire was an authority on marsupials and monotremes, having completed considerable work on the wombats brought back from the Baudin expedition by François Péron, but the lack of access to koala specimens continued to plague the European scientists. Even the great anatomist Georges Cuvier was stymied by the koala. The koala was one of only five mammals he chose to illustrate for his multivolume *Animal Kingdom*. Refusing to reproduce Lewin's koala, he had his own drawn, which looked more like a lynx with strange, clawed hands than a koala. When the book was translated, the English publishers carefully replaced his 'koala cat' with Lewin's image, but the lynx-like representation stayed in subsequent French editions.

A multitude of new names, descriptions and drawings were published in the early nineteenth century, but none was based on any additional evidence or analysis of the animal – dead or alive. At a time when fearsome debates raged in Europe over possums and platypus, accounts of koalas were brushed off with the flippant comment, 'It can be distinguished by the lack of a tail.' Eventually, the balance of opinion settled on a compromise between Blainville's genus *Phascolarctos* and the German naturalist Georg August Goldfuss's species name *cinereus* of 1817. After this, scientific interest in the koala evaporated. Not even the resolute scientists of successive French expeditions, who contributed so much to early Australian biology, managed to acquire a koala until the 1840s.

Charles Darwin, visiting New South Wales in 1836 on his famed *Beagle* voyage, was charmed to see (and shoot) a platypus in its natural habitat, but did not even mention the koala. One of the sites Darwin made a particular point of seeing was Govetts Leap, a spectacular lookout over the great valleys and sandstone cliffs, shrouded in the blue eucalypt mist that gives the Blue Mountains their name. It's a shame Darwin didn't meet the namesake of the lookout, the surveyor William Romaine Govett, as he had just published one of the first detailed accounts of koalas, in the London *Saturday Magazine*.

Govett was presumably not familiar with the scientific publications declaring the koala's place in the great taxonomy of life. He realised that they were neither bears nor monkeys and, after originally thinking they were like sloths, eventually concluded they were more like the lorises or lemurs of India. He had shot several and had others caught by his Indigenous companions, keeping the animals for a time in his camp to observe them.

He noted their thick blue-grey fur, naked hands and round, dark eyes that were 'sometimes expressive and interesting', as well as their melancholy cry, which, he said, 'excited pity'.

'The countenance is by no means disagreeable, but harmless and pitiful,' Govett wrote. 'They are certainly formed differently from every other species of the quadrumana, and it is probable they possess different enjoyments. They are exceedingly inoffensive and gentle in manners, if not irritated.'

Govett's pragmatic yet sympathetic account of the koala remained one of the few early European accounts of the animal for some time. As colonists spread out across the far reaches of the continent, curiosity about koalas remained patchy and sparse. Australia's intriguing wildlife remained of little intrinsic interest to most of the country's newest inhabitants – a nascent nation of farmers and traders – until someone discovered a way of making money from it.

17

War and Guns

The koala is perched between two men, who are cross-legged on the ground in their uniforms. Emu plumes adorn the slouch hats of the 5th Light Horse Regiment, which comprises volunteers from across Queensland. The men had discovered the young animal in a tree while on a route march to Sandgate in Brisbane. One of the men swiftly climbed the tree and grabbed the young koala, bringing it down to the ground.

In the photograph, they are outside a tent as the men await deployment to the Middle East. Some accounts claim that the koala went with them to Palestine, but I can find no further information about it other than the picture published in *The Queenslander* in December 1914. I hope they released it before they left, but it's unlikely. Animal mascots were popular with troops embarking for war, and what could bring them better luck than an iconic Australian koala? The Victorian koala mascot of the 3rd Divisional Supply Column, Teddy, did not survive the journey to England, but another koala features in army photos taken at a hospital in Cairo.

The battlefields of the Middle East were difficult enough for men and boys to survive, much less a koala. But even if the koala stayed in its forest, it would not have been much safer. The war on the home front would prove just as brutal and bloody as the one in Europe.

On 1 January 1901, the six former British colonies of Western Australia, South Australia, Tasmania, Victoria, New South Wales and Queensland officially joined forces in a national federation of Australia. In theory at least, the country became a single political entity, allowing for uniform national laws governing trade, defence and immigration and for the beginning of a slow and protracted unshackling from the influence of British imperialism.

The drive for federation consolidated a surge of patriotism and national pride that was to become the hallmark of the early twentieth century. Australians had already begun to represent themselves as a nation on the sporting fields and the fields of war. And Australian plants and animals became the motifs for this movement. Gum trees and lyrebirds, wattles and emus – the distinctive and ancient Australian fauna were seen as emblematic of the new commonwealth, as European colonisers sought to establish cultural connections to the land they had annexed. Cricketers and soldiers were depicted as kangaroos. Australian animals, like the koala, became and have remained a far more recognisable symbol of the country than any colonial flag.

In 1904, Norman Lindsay published his first sketches of a koala, who soon became 'Billy Bluegum', illustrating a satirical story about his efforts to civilise 'barbarian bush bears'. The character of Billy Bluegum was recruited to many national campaigns, including during both world wars in 'Billy Bluegum takes to the Gun tree' and on the winners' podium of the 1956 Melbourne Olympics. In his classic children's book *The Magic Pudding*, first published in 1918, Lindsay re-imagined his koala character as Bunyip Bluegum, one of the story's protagonists.

In 1914, many Australian troops went to war adorned with the pelage of Australian fauna. As well as emu plumes, the Light Horse occasionally wore a band of wallaby fur on their slouch hats. Hundreds of products, from flour to brass beds, were branded as 'koala' – testimony to the growing popularity of this small animal.

This popularity piggybacked onto another burgeoning international phenomenon – that of the 'teddy' bear. By strange coincidence, two manufacturers on opposite sides of the Atlantic created stuffed toy bears in 1902, both intended for the US market. The fate of the German toy bears is unknown, leading to the suggestion that they were shipwrecked on the way to the United States. But the American version became a sensation, and was sold as the 'teddy' bear after President Theodore 'Teddy' Roosevelt, who had refused to shoot a black bear which had been chased, clubbed and tied up for the purpose. Roosevelt, an enthusiastic big-game hunter, was almost certainly motivated by pride, rather than empathy for the bear.

Teddy bears went on to become a favourite children's toy around the world, gradually morphing into progressively cuter forms, with larger eyes and foreheads and flatter snouts. Their popularity was further entrenched in widely loved children's stories like *Winnie-the-Pooh* in 1926 – 'A bear with very little brain' – and in 1958 *A Bear Called Paddington*, who appears with the engaging note attached to his jacket: 'Please look after this bear.'

Koalas are surely more 'teddy bear' than a true bear. They need no modifications to their wide face, forward-facing eyes and clinging posture that elicits such a strong parental response in humans. By the 1920s, toy koalas – with movable limbs, button eyes and a big rubber nose – were being manufactured in Australia and overseas and shipped all around the world. Small children everywhere could cuddle up to these soft toy companions that were 'wrapped in a rabbit skin' – or, more commonly, mohair, kangaroo, wallaby and even koala fur itself.

In a tragic irony, the abundance of toy koalas was enabled by the ready availability of their furry pelts for coats, muffs, purses and handbags. Far from being symbolic of our affection for this animal, their transformation into cuddly toys was a side effect of the heavily commercial, mass-market trading in the slaughter of hundreds of thousands of koalas in the wild.

Native animals became a major source of both food and sport for the settlers as the encroaching colony spread. Kangaroo hunts were common sport for imported 'greyhounds', soon bred into lanky, powerful 'kangaroo dogs'. The impact was noticeable even by 1836.

'A few years since this country abounded with wild animals,' noted Charles Darwin, invited to participate in a rather unsuccessful kangaroo hunt, 'but now the emu is banished to a long distance, and the kangaroo is become scarce.'

It's unlikely that koalas were eaten much by European settlers, given that they barely seemed to notice them. In the 1850s, a hunter on the Mornington Peninsula in Victoria commented that their flesh was 'not unlike that of the northern bear in taste', although whether that was a good thing was not entirely clear. Koalas may not have been valued for food, but the demand for all kinds of fur in America and Europe was huge. Koala pelts stretched into neat rectangles of thick, soft fur, making durable rugs and coats which were prized for being waterproof.

The common methods for catching koalas involved either poison or wire snares. They were rarely shot, prompting the suggestion that koalas were not worth wasting bullets on. In fact, koala hides had a reputation for being so tough that bullets would bounce off them. Melbourne game-hunter H.W. Wheelwright noted that they were 'extremely difficult to shoot on account of their thick hide', in addition to their thick coat, small size and habit of hiding behind branches. More importantly, though, shooting risked damaging the pelt.

Instead, the animals fell victim to much more cruel and inhumane methods. A jam tin of water dosed with cyanide would be placed near the foot of a tree or a hollow log. Any animal unwary enough to drink the water during the night suffered a painful, agonising death. The technique for snares was not much better.

'Trappers place slanting saplings against the likely tree, and arrange on each the deadly wire noose through which the "possum" will thrust his head coming down,' reported Nora Howlett, who summarised the horrors on the 1927 hunting season in the late 1970s. 'In the early morning, before dingoes and crows have disturbed the corpses, the trapper does his rounds to collect the strangled "possums" and bears.'

Young joeys were thrown to the dogs, while older ones might be released, for later capture, to take their chances on their own.

It's hard to know the exact number of koalas hunted for fur – for either the local or overseas markets – as koala fur was often sold out of season as 'wombat' and many different Australian furs were sold in the United States under the general label of 'opossum'. But demand for Australian fur was high. Annually, around 10,000 to 30,000 pelts were exported to London, while in 1889 an astonishing 300,000 pelts were sent. By the early 1900s, the koala population had plummeted in Victoria, New South Wales and Queensland.

Concerns over their fate led to them being declared a protected species in 1898 in New South Wales, but 57,933 koala skins were still exported from Sydney in 1908. The southern populations were all but wiped out, first in South Australia and then in Victoria, despite protective legislation.

Hunting continued at a ferocious pace in Queensland. There were open seasons on the animals in 1915, 1917 and 1919. The 1919 season, from 1 April to 30 September, yielded 1 million koala skins. Nationally, 205,679 koalas were reported as having been killed in 1920 and 1921 for their fur. In response to public outrage, in 1921 Queensland passed the *Animals and Birds Act* to protect koalas, and in the same year Commonwealth legislation was introduced to control the export of koala skins. However, this did little to abate the massacre.

In 1924, 2 million koala pelts were exported from the east coast. In 1927, the Great Depression led the Queensland government to offer a limited open season of one month for koala hunting.

'From all parts of the State voices are raised in protest against the action of the Government in proclaiming an open season of one month for koala bears,' declared the *Daily Mercury*.

But the protests did not work. Before Queensland's final official season in August 1927, 10,000 koala-hunting licenses were issued to boost rural employment and reduce 'uncontrolled' koala numbers. Within a few days of opening, over 70,000 skins were sold in Brisbane markets. During 'Black August', 600,000 koalas were killed that month across the state. Photographs show koala skins stacked into overflowing piles on warehouse floors, and 3600 pelts loaded onto a trailer by one group of hunters in Claremont, Queensland.

Demand was high for the soft, thick fur of koalas in the coat, hat and glove trade in the United States. Most of the furs were sent to St Louis in Missouri.

'It is hunted and ruthlessly destroyed for the sake of vain women with more money than brains or the milk of human kindness,' a

Sydney journalist declared, 'in order to make for them fashionable coats and other forms of wearing apparel.'

David Stead, the president of the Wildlife Preservation Society of Australia, railed against the 'death knell' of the koala, predicting that the cruel and senseless slaughter of Australia's most-loved animal for a few paltry sovereigns would earn the scorn of future generations.

The documented number of deaths is almost certainly an underestimation. This tally is higher than the number of koalas that have ever been counted as living in Australia. The true natural abundance of this species is tragically only indicated by this horrifying death toll.

Nor did the killings stop. Despite public outcry after Black August, hunting continued in the closed seasons, often under the guise of obtaining 'wombat fur'. In the end, it was not the Australian government who stopped the slaughter, but an American president.

In 1930, David Stead appealed directly to President Hoover, who had once worked in the goldfields of Western Australia. Hoover responded by prohibiting the importation of koala and wombat skins into the United States, and the trade eventually dried up.

Three years later, the Australian government finally banned the export of koalas and koala products.

18

Saving the Koala

The weatherboard cottage was small, basic and remote. No telephone or mail delivery, no sewerage or window screens. It had an outside loo, a wood stove, a rainwater tank and hardly any neighbours. Warrimoo in the Blue Mountains was barely even a town – just a cluster of houses around a railway station. The only significant building was the narrow two-storey general store–cum–post office, which looked like it belonged in a crowded Sydney high street, rather than the open expanses of the country. But it was surrounded with trees and wildlife. The bush track down the end of Florabella Street disappeared into a bird-filled forest of angophoras, banksias and stringybarks that led to the fern-filled gully of Glenbrook Creek and the much-famed Warrimoo rock pool.

I have always imagined the author and illustrator Dorothy Wall and her young son, Peter, living in their little rented house in the Blue Mountains wilderness, listening to the calls of the koalas at night, or pausing to watch as one crossed their path as they walked through the bush. I assumed this must have been the source of Dorothy's famous 1933 book, *Blinky Bill*, about the adventures of a feisty little koala.

But then, as now, there were very few koalas in the Blue Mountains. Indeed, as far as Dorothy was concerned, there were hardly any koalas in New South Wales at all. In the story, Blinky Bill and his mother come from Gippsland, in eastern Victoria, and they travel up to New South Wales after Blinky's father is shot by a hunter at the beginning of the book. Here they meet local resident Mrs Grunty, who explains that, 'In New South Wales I think we could wander for miles from one corner to another and never meet a bear.'

In fact, the direct inspiration for Blinky Bill came from a koala who performed on stage with Brooke Nicholls and his radio star kookaburra, Jacko, which punctuated the show with its inimitable chortling laughter. Dorothy had been asked to illustrate a book about Jacko, and no doubt used her time to also observe and draw the quiet koala. Years later, when writing her Blinky Bill series, Dorothy had to visit the zoo in the city for models, and was apparently given special permission to enter the koala enclosure.

Blinky Bill is a classic coming-of-age story about the adventures of a young boy in the bush, drawing as much upon Dorothy's son Peter as it does on koalas themselves. But the narrative contains a powerful conservation message – about the threats to the Australian bush and its animals. Starting with their escape from hunters in Victoria, then through the emptied forests of New South Wales, their encounters with humans and eventual arrival in Taronga Zoo, *Blinky Bill* not only gave voice to growing public concern over the fate of the koalas, but also inspired generations of Australian children with a great passion for their local wildlife. Nor could Dorothy leave her characters to their fate in captive confinement. Her final book sees them escape back to the bush. Her books reflected a growing concern about the fate of many Australian animals, and particularly koalas. As one reviewer put it:

No animal seems to supplicate more strongly for protection. Yet they are wantonly slaughtered each year in their thousands for skins. *Blinky Bill* will do much to endear these animals to children, and, it is hoped, do something to curtail, if not altogether abolish, the annual slaughter which must eventually exterminate them.

It was not just Australians who were increasingly fascinated by koalas, though. They were in great demand overseas as well, particularly from international zoos. But getting them there was not easy.

The first koala known to have survived the long ocean crossing to Europe arrived in London in 1880 – sustained, apparently, by dried eucalyptus leaves. She had been purchased by the Zoological Society of London, which operated the London Zoo, and survived for fourteen months on a diet of dried leaves and fresh ones shipped from Australia. Having spent her life as a pet, the young female was kept in the keeper's office, resulting in tragedy when, unattended overnight, she got her head stuck under the heavy lid of the washstand and died.

It's a heartbreaking story and the singular affection many keepers have for these animals must only have made their inevitable deaths even worse. Keeping animals safe and well in captivity is always challenging, fraught with failure and tragedy despite the best of intentions. The challenge for koalas was even more pronounced. I can't imagine a koala, with limited reserves of body fat, making do with dried gum leaves or even ones 'fresh' from a month-long voyage from Australia. It's possible some of them may have survived because they were so young that they were still being fed milk. In total, six koalas arrived at the London Zoo between 1872 and 1931 – four in the late 1800s and another two between 1922 and 1931 – none of which survived more than a year. Remarkably, the koalas fared worse than the nine (now extinct) thylacines sent to London, two of which survived for five and eight years respectively.

Today, koalas are only transported on direct flights with no stop-overs, with daily access to fresh-cut leaves. The animals that survived such long voyages in the past may well have arrived in a half-starved and perilous state of health, unbeknown to their owners. It's no surprise that long-term survival was rare.

The list of foods that people have tried to sustain koalas on is horrifying: tea-leaves, peppermints, cough drops, sweets, fruit, nuts, bread, milk, whisky and tobacco. It would be many years before koalas were successfully transported overseas and survived, and even today they remain one of the most expensive and difficult of captive animals to care for away from their home forests.

The ban on koala exports in 1933 slowed down efforts to send koalas overseas. Australian zoos were wary of allowing animals with such particular diets to travel without assurances that the right food would be available for them.

Eucalypts, however, have no such difficulties with moving over-seas. In fact, they have been one of Australia's most successful reverse colonists – populating every habitable continent of the globe, some-times in pest proportions. Southern blue gums in particular are one of the most widespread hardwoods in the world, with vast plantations grown largely for paper production. In places like southern California, Australian eucalypts have become widely naturalised. During the late 1800s, eucalypts were established for timber, firewood and windbreaks. San Diego town planners planted river red gums along the city streets for their shade, beauty and speed of growth. Plantations of southern blue gums sprang up as a cash crop. If koalas were going to do well any-where outside Australia, this was the perfect place to try.

The San Diego Zoo already had an abundance of mature euca-lypts across their grounds when they received their first koalas from

Sydney's Taronga Zoo in 1925. Not everyone had this advantage and even with it koalas remained a challenging species to keep. Koalas sent to other zoos sometimes inexplicably failed to thrive, with eucalypts or not.

All too often, koalas would arrive and simply refuse to eat the leaves provided. Or eat them for a while and then refuse them later. Or eat them and fail to thrive anyway. Some overseas koalas had to be sustained on gum branches flown in over thousands of miles every day. Clearly there was something wrong, something the koalas could tell about the leaves that no-one else could. It's a problem that keepers still face today.

'We have to provide them with a range of species each day,' explains Ash, one of the koala keepers at Cleland Wildlife Park. 'Might be that one of the trees is no good for some reason, that they don't like it, so we have to make sure there is something else for them to eat. You just have to trust them to know what they need.'

By 1939, the extinction of koalas on mainland Australia seemed both inevitable and imminent. Following what seemed like an almost systematic effort to wipe out the last stronghold of the species in Queensland in 1927, koala populations there were largely reduced to only those in the arid inland forests. In addition to hunting, a series of catastrophic wildfires through the late 1800s and early 1900s had a massive impact on koala populations in Victorian forests. Five million hectares of forest and farmland were burnt in the Black Thursday fires of 1851, and a further 260,000 hectares went in the 1898 Red Tuesday fires. The koala death toll rose further with 2 million hectares destroyed during bushfires in 1926 and 1939, with the loss of 131 human lives and an incalculable amount of wildlife. Forest fires in the south-east of South Australia in the 1930s reduced populations of koalas at the most westerly edge of their range to undetectable levels.

Koalas were widely declared extinct – in South Australia, New South Wales and mainland Victoria – although whether they actually were is uncertain. Without doubt, the remaining animals were exceedingly scarce, widely dispersed and in small, isolated pockets of remnant vegetation. For example, a broad community survey of koalas in Victoria in the 1920s reported only 500–1000 animals across the whole state, including a small distinct population in the Strzelecki Ranges.

Local residents, though, witnessing these declines, had already started taking koala conservation into their own hands. It's thought that one local resident, in particular, initiated this action. Jim Peters was a farmer, fisherman and keen naturalist, who collected one of the first specimens of the extremely rare Leadbeater's possum. It seems likely that during the 1898 fires, Jim rescued a handful of koalas, bundled them into a small boat and transported them to safety on French Island. The island had a healthy supply of manna gums, but no resident koalas.

The koalas enjoyed the move. By the 1920s, naturalists counted over 2000 koalas in an 8-kilometre stretch along the west coast of the island, and the manna gums they lived in were being stripped of their leaves and some were even dying. The koalas were quite literally eating themselves out of house and home.

Local farmers requested the animals be culled, but instead the government offered two shillings and sixpence for each live bagged koala delivered to the local jetty for translocation. In 1923, some fifty koalas were sent to nearby Phillip Island and six animals to a fauna reserve on Kangaroo Island, set up by the curator of mammals at the South Australian Museum, followed by twelve more in 1925.

In 1934, 165 koalas were moved from French Island to Quail Island, a small reserve of 3000 acres in the north of Westernport Bay in Victoria, forested with manna gums. By 1943, the koala population had boomed to between 650 and 1000 koalas, and visitors reported that 'many trees were dead, a few had a little foliage on inaccessible tips, the rest were bone-bare and seemed to be dying'.

It wasn't just the trees that were dying either. One regular visitor to Quail Island, Ronald Munro, was horrified by what he witnessed as the koala population continued to increase.

'Hundreds of starving bears, many with little babies clinging to their backs, were sitting up dead trees or slowly moving about in search of food,' he said. 'With their thick woolly fur they still appear well and healthy from a distance, but all those I examined closer were just skin and bone, and the babies were skinny weak little creatures.'

In 1944, a film of Quail Island revealed the dying koalas and deteriorating conditions. It was shown in local cinemas, and articles were published with photos of dead koalas beneath stripped-bare trees.

The ensuing public outcry called for some of the animals to be translocated to the mainland, as had been done for French Island, but the authorities remained unconvinced, declaring that the film gave the mistaken impression that the koalas were starving and that it was a 'common sight to see koalas sunning themselves on dead timber'. Wartime censorship laws were swiftly enacted to ensure that the film was not screened overseas.

Eventually, however, translocations did take place, and over the next four decades some 14,000 koalas were moved from the islands of Westernport Bay to throughout Victoria. These animals, particularly those from French Island, swiftly expanded back into their old ranges across Victoria, so successfully that sterilisation and contraception programs would eventually become part of koala management in the state. Other animals were released back into areas that they had not occupied for millennia, like Kangaroo Island and the Mount Lofty Ranges.

The eighteen French Island animals released on Kangaroo Island in the 1920s settled in well and soon expanded in number. By 1937 there were reports of 100 koalas on the island, which were said to be even larger individuals than their founders. The use of Kangaroo Island as a refugium for endangered mainland species, like the koala and also the platypus, was strongly supported. Even my great-grandfather apparently thought so. I am startled to come across a newspaper article describing the then Minister for Agriculture in South Australia, Percy Blesing, as an advocate for the protection of Australian wildlife, particularly koalas.

'They are very beautiful animals and harmless in every way, as they live almost wholly on gum leaves,' he said. 'The Kangaroo Island situation was ideal, as there are no foxes, rabbits or other pests there.'

By the 1940s, more koalas had arrived on Kangaroo Island, this time along the Willson River. They came from the koala farm in the city and the breeding farm at Williamstown in the Adelaide Hills, where they had been bred very successfully. The koala farm's director, Keith Minchin, planted additional red gums, white gums and South Australian blue gums along the Willson's banks for the koalas' future enjoyment. A few locals objected, apparently worried the animals might get into their fruit orchards, but Minchin assured them that they would stick to their preferred gum trees along a 15-kilometre stretch of the river.

And for a time, he was right. Up until then, the koalas appeared to have kept to the areas near where they were released. My cousin Jeremy Leech was a student in the late 1950s, and he remembers helping George Lonzar and Slim Bauer on a project that tracked koalas on the western end of the island. Even though they had been on the island since the 1930s, they seemed at that time to have remained fairly still, restricted to a small patch of manna gums near Rocky River in Flinders Chase National Park.

'When we tracked them, though, we found that each night, the koalas would leave their manna gum clump and travel out in a wide arc before travelling back home,' Jerry tells me. 'It seemed like they were looking for new sources of food, but not finding any within a night's search.'

Eventually, though, they did. As soon as the koalas found another patch of manna gums, they stayed in that area, which also became a staging post for future expeditions. From here, the koalas would launch their expansion out across Kangaroo Island, locating new manna gums and other favoured browsing trees.

Rough-barked manna gums are quite widespread across Kangaroo Island, growing in the damper gullies. By the 1940s, they were showing signs of over-browsing, with the trees dying out in some areas altogether. By 1994, the problem had become dire. Most manna gums on the island were severely depleted, with half their canopy missing, and in 1997 the South Australian environment department took action.

Public concern for koalas was high. School children planted trees for them on the mainland, hoping some would be brought to their area. But there were few parts of South Australia that had enough trees to support koalas. Instead, they were relocated to the south-east of the state, where they had once been well established, and to the still forested areas of the Mount Lofty Ranges, joining the pre-existing population. Following the Victorian example, a sterilisation program began which successfully slowed the koalas' population growth and reduced their density from three koalas per hectare to one.

Around the same time, though, a new range of gum trees was planted on Kangaroo Island. In 1997, in an effort to stimulate regional economies, the South Australian government began subsidising the planting of southern blue gums, and this support continued into the 2000s. Large areas of the trees were planted on the island. As the blue gums matured, the koalas moved in to take advantage of this added bounty. Very soon, these private plantations began supporting an even larger number of koalas. By 2015, the estimated population on just 200 hectares of Kangaroo Island had reached over 48,000 koalas.

From just a small handful of animals, rescued from the brink of extinction, the koala population across southern Australia has rebounded, to over half a million animals.

On the face of it, the recovery of the koala from near-extinction is a rare and remarkable conservation success story. The process of rescuing critically endangered populations, establishing them on islands or reserves, safe from major threats, and breeding them sufficiently to successfully re-establish them in their original mainland habitats is a last-resort strategy and notoriously difficult to achieve. But it is also the gold standard of conservation programs across the world. This technique was used to save Przewalski's horse and the American bison, as well as countless birds, reptile and frog species, particularly those from isolated islands vulnerable to species extinctions, like New Zealand, Mauritius, Madagascar and the many Polynesian islands across the Pacific.

Despite being a continent, Australia too is one of those vulnerable islands. Its great diversity of species, gifted by the country's size, combined with its long and isolated evolutionary history and relatively recent colonisation, has resulted in one of the worst extinction rates in modern times. Nor has Australia been particularly successful at slowing or reversing this trend – with the single dramatic exception of the southern koalas.

So, why is it that the story of the remarkable recovery of the koalas across the southern part of their range, rescuing one of Australia's most distinctive and unique animals, is not celebrated as a great conservation success, but instead seems to be widely regarded as yet another environmental catastrophe?

VI
FUTURE TENSE

Dry thunder rumbles across the mountains, like brooding giants slowly wakening their aches and pains from overly long sleep. Dark clouds drift out to sea, refusing to relinquish their precious cargo. Forked tongues of light split the purple sky, seeking out the highest point to discharge their overload of energy. An explosion ricochets across the forest as the tallest, oldest tree is hit, its dead heart igniting, scorch marks shrieking down the pale stem to disappear in trails of smoke into leaf litter.

Thin grey smoke puffs and curls, thickening long trailing spirals upward that drift sideways, like ghostly curtains, between the trees. A gust of wind glows red beneath the leaf litter. The smoke thickens. Heat radiates, ignites in golden figures that dance and sway in the air, creeping, spreading through the long dry grass, warming, growing, engulfing shrubs and bushes, expanding, overwhelming. The flames caper and frolic across the forest floor, leaping to low branches, flinging themselves up the dry bark, nibbling and biting at leaves. The heat rises centigrade by centigrade: 50 degrees, 100 degrees, 500 degrees, 1000 degrees. The world combusts in roaring fury. The forest explodes in brightness and is overcome by dark.

Fire burns the forest, burns the earth, burns the air. Flames flick into the canopy, erupting in an explosion of gases. Ancient trees groan and die, their material forms converting back into the gases they came from. They burn on a funeral pyre of their own creation.

The forest is still. Nothing moves but a single leaf tugged by a stray breeze drifting in an otherwise airless sky. An eagle returns, circling the sun on giant's wings, wheeling on the thermals curling from the still-warm earth. The forest below is dark, colourless, despite the bright day. Heat and dry seep into dead timbers, cracked grey and white.

Only the woody nuts move, spitting a million tiny life-support rafts into the atmosphere – the trees' last act – released to their uncertain destiny. ⌇

19

Sex, Disease and Genetic Diversity

It may well have been a small tropical mouse that started the problem. Melomys or mosaic-tailed mice seem innocuous enough: little fawn-brown rodents, native to the wet tropical islands of Indonesia and New Guinea. Large eyes, long whiskers, hairless tail. When lower sea levels 5–6 million years ago brought the Australian and Asian landmasses closer together, these small mice found themselves clinging to debris swept down the turbulent Fly and Sepik rivers, out to sea and then onto offshore islands.

Some of these mice remained stranded on the islands. The Bramble Cay melomys survived on a low sandy atoll in the middle of the Torres Strait, with only seasonally nesting seabirds and turtles for company, until the last of the mice died on the increasing waves of slowly rising waters. It was the first documented Australian mammal extinction due to climate change.

The grassland melomys, however, made it to mainland Australia, making a new home for itself in the wet eucalypt forests and grasslands of northern Queensland, doing no harm to itself or anyone else, apparently, until scientists found its genetic fingerprints on the smoking gun of a deadly koala retrovirus. Just as rats and their fleas have carried the bacterial disease known simply as 'The Plague' through human populations, killing millions (up to half the world's population, by some estimates) in successive waves from the fourteenth century onwards, this little melomys would inadvertently carry death and disease to Australia's koalas – their very own version of AIDS.

The mice were not the only invaders to bring diseases with them.

We've learnt a lot about epidemics and pandemics in recent years. Not just scientists, but the public too. COVID-19 has made us all experts, it seems, on quarantine and vaccination, on how diseases spread and how to curtail them. We are experts, too, on failing at these things – we've realised we are slow learners, even after centuries of plagues, poxes and influenzas have successively swept through our populations.

We have learnt, if not always accepted, how these diseases are transmitted and the strategies to minimise their spread and impact. Pandemics and infectious diseases have played a major role in human populations. Our mobility, close living quarters and desire to share them with other species have all contributed greatly to the rise and spread of new diseases, which have often had a devastating impact on the global population. Human pandemics are nearly always associated with movement and change. Wars and colonisation have enabled the rapid spread of diseases. So too has air and sea travel and unexpected climatic changes. But the origin very often lies at the interface between humans and other animals. Nearly all our major pandemics have originated from close contact, inadvertent or intentional, with other species, allowing the transmission of a sometimes-innocuous pathogen from one species into us, often in a more virulent form. Whether with pets, domestic stock or wildlife, our interactions with other species have driven much of our infectious disease burden – from influenza to AIDS to COVID-19.

When Europeans first colonised Australia, they brought with them a huge range of new diseases that decimated Indigenous communities. Smallpox, syphilis, tuberculosis, influenza and measles all radiated swiftly across the country as people fled from the infections, killing up to 70 per cent of the local population, like a covert military force clearing the country prior to settlement.

Colonisation brought devastating epidemics to Australian wildlife as well – including koalas. Even at the turn of the nineteenth century, it was apparent that koala epidemics were having a major impact.

'The first drastic shrinkage of range apparently resulted from epidemics of some form of ophthalmic disease and periostitis of the skull introduced as a result of settlement,' reported Ellis Troughton, 'which is said to have swept away millions of koalas in the years 1887–1889 and 1900–1903.'

Although Troughton provides no source or details for these observations, it seems likely that the symptoms of ophthalmic disease resulted from chlamydia bacteria, while the skull deformations are consistent with a form of koala AIDS caused by the retrovirus KoRV, both of which continue to decimate the east-coast populations of koalas today.

Like human AIDS, the koala retrovirus suppresses the immune system, reducing the koala's ability to fight off illness. They are multiplier diseases. As well as causing blood and bone marrow diseases, lymphomas, leukaemia and craniofacial tumours, these retroviruses also increase koalas' susceptibility to chlamydia.

Viruses are simple yet formidable parasites. They are little more than weaponised genetic code wrapped in a protein coat, less than

500 nanometres in size. Too small to be seen under a light micro-scope, they insert themselves into the living cells of their host, hijack a cell's command system and force it to make copies of the virus instead of performing its normal functions. Safely sheltered inside the host cells, viruses like Ebola, herpes and coronavirus are difficult to treat because the body's immune system doesn't recognise its own cells as invaders. Antiviral drugs can't attack the virus directly – they can only block the virus's ability to bind with and infect cells and replicate. Once you have a herpes virus, you are stuck with it forever. All you can do is treat the cold sore and prevent the virus from spreading.

Retroviruses, like HIV, are even more insidious. They not only take over activity in the host's cell, but they physically insert them-selves into the host's DNA, becoming impossible to eradicate and even more difficult to treat. They don't just embed themselves once either – they create multiple copies of their genes all through the host's genetic code, causing chaos in the cell's functions. Currently, HIV can only be suppressed to undetectable levels by a constant reg-imen of anti-viral drugs, but it never disappears entirely.

If that wasn't enough, the koala retroviruses have added a whole new level of nastiness to their arsenal. In addition to the body cells, some variants of KoRV can also infect the koala's reproductive cells – the eggs and sperm. So, it passes not only from one koala to another through contact, but also from mother to child during birth. In north-ern koala populations the KoRV is often preloaded into the offspring's DNA; joeys are born with a full complement of the retrovirus already infecting every single cell in their bodies.

Retroviruses are not new. Even hominid DNA shows signs of hav-ing been invaded by a similar retrovirus about 3 million years ago – the legacy of which we still carry in our cells today. About 8 per cent of our

DNA is thought to be from retroviruses. They have also been present in the koala genome for millions of years, with several variants. The koala retroviruses share many similarities with a retrovirus that causes leukaemia in gibbons in South-East Asia, which was probably carried to Australia by a melomys mouse or perhaps a flying fox. But new and more deadly varieties are infecting the koala population today, causing some scientists to suggest that the animals are undergoing a 'genomic invasion' as the koala retrovirus spreads from the heavily infected northern koala populations to the southern populations with relatively little disease.

Retroviruses may be using a koala's own genes against it, but koalas also take advantage of the retroviruses' ability to produce new mutations. It's a classic co-evolutionary war – an escalating arms race – that is going on in the east coast forests. Small wonder the koala landscape sometimes seems like a battlefield. But there are signs that the koalas might be fighting back. Scientists have observed for the first time a new kind of immune response to retroviruses from a secondary immune system. This secondary system uses molecules called piRNA, instead of antibodies, to identify and target viral DNA. Like the primary immune system's production of antibodies, it usually takes a few days for the body to gear up and produce enough piRNA to mount a defence against the new invader. But researchers have recently found that koalas also have an immediate innate response to KoRV. By recognising a common feature of all retroviral DNA, they are able to launch a quick-attack force of piRNA that reduces the impact of the invasion while their systems ready their main genome immune response. Instead of dying straight-away, the koalas only fall sick, giving them a fighting chance of recovery.

It's not in the best interests of any virus to kill its host outright. If the host population dies out completely, so too does the virus. There's an adaptive advantage to milder strains of disease – natural vaccines, if you like – that train the host's immune system to overcome more virulent and deadly strains. Over time, viruses attenuate and less deadly strains become more common.

In the war between koalas and retroviruses, neither side can win. The battles are short and brutal, with periods of uneasy peace where both species reach an accommodation and learn to live together.

Despite the horrible diseases and deaths that they cause, retroviruses may not be all bad news, though. In a way they also represent a kind of evolutionary supercharge. Species acquire the variation they need to adapt and survive in a changing world through genetic mutations, which happens whenever a new individual is created with every new generation. Creatures that reproduce rapidly (like bacteria) acquire genetic diversity more quickly than those that breed more slowly (like mammals). But mammals have another way of creating genetic diversity. By reproducing sexually, they mix their genetic material at each generation (rather than producing clones), which massively increases variety, speeding up evolutionary processes. By inserting themselves throughout the genetic code, and particularly in the inherited reproductive cells, retroviruses function in a similar way.

'Evolution is driven by mutation,' says Professor William Theurkauf from the University of Massachusetts, who studies how koalas' immune systems respond to retroviral invasions. 'What these viral invasions do is stir the genetic pot, they basically generate a lot of genetic diversity.'

What's more, retroviral invasions also ramp up natural selection by increasing mortality. What doesn't kill you makes you stronger, as they say.

The fortuitous selection of animals largely free from chlamydia and KoRV for the French Island population and their resulting expansion through the Victorian and South Australian forests has almost certainly saved the southern koala from approaching extinction. Is there a downside to this fast-breeding population of large, robust koalas? Genetic diversity, which accumulates over generations, is about evolutionary potential, a long-term insurance policy that theoretically enables animals to adapt to new diseases or environmental changes. The small size of various founding populations of translocated koalas has been replicated over a number of sites, reducing genetic variation and increasing the risks of inbreeding. Genetic diversity in human populations, for example, decreases with distance from the evolutionary origin of our species in East Africa, presumably as an effect of being descended from a small group of founders. Similarly, each time a small number of koalas starts a new population, this founder effect is magnified. It sounds like a recipe for a genetic disaster.

Not that there is much evidence of maladaptation in any of the southern koala populations currently. Some populations of southern koalas have a greater genetic risk of kidney disease, which makes them more vulnerable in hot, dry weather, often causing them to approach people for water. Koalas on Kangaroo Island have slightly higher rates of testicular abnormalities but neither of these traits have prevented these populations from thriving. Even so, this theoretical future risk of genetic inbreeding has led some key organisations to refuse to recognise the legitimacy of southern koalas as viable koalas at all.

While the South Australian and Victorian governments have claimed populations over 200,000 and 450,000 southern koalas in each state respectively, one of the peak conservation groups claims that Victoria has a maximum of only 23,000 animals and South

Australia only 13,000 animals. The Kangaroo Island population is entirely excluded from these calculations.

This is not because it disputes how koalas are counted, but which koalas should be counted. The Australian Koala Foundation acknowledges that there are said to be hundreds of thousands of southern koalas, but it argues that while there 'may be some isolated habitats and islands that have large numbers (sometimes in the 1000s), we believe these populations cannot be considered to have long-term viability because of their inbred status'.

But just how inbred are these koalas? And how much does it matter? Studies have revealed more genetic complexity than expected. Koalas in the Adelaide Hills, for example, are often assumed to be descendants of a series of bottlenecks on Kangaroo Island and French Island. In fact, they have a much more diverse ancestry, being genetically connected with animals in Queensland, New South Wales and Victoria. The unregulated transportation of koalas meant that they were often moved to new places as pets, as wedding gifts, for display or for amusement. In South Australia, they were reported to have sat on the bar at the Eudunda Hotel, been at a garage in Enfield, in the Koala Farm holding pens at Williamstown, on doctors' properties in the Adelaide Hills, and even displayed in a store in the city before being sold. Some of them survived, escaped or were released intentionally. One was even noted to have been present when the new Chrysler factory was built at the industrial Tonsley Park in 1964, well before the arrival of koalas from Kangaroo Island.

Genetic diversity is lower in the Victorian and South Australian populations than those in New South Wales and Queensland, although whether this reduction is significant is unknown and many smaller northern populations are now also genetically isolated and inbred. Long-term species survival is a lottery – a diverse gene pool might increase the odds, but it's no guarantee. The descendants of the French Island animals seem to have hit the jackpot genetically, while the genetically more diverse northern koalas are struggling with disease and habitat loss.

The biggest impact on koala's genetic diversity seems not to have been recent translocations or the slaughter of millions of animals in the early twentieth century. Comparisons of nineteenth- and twentieth-century museum skins with living koalas have revealed that koala genetic diversity was low long before European colonisation. In fact, it seems to have been the Last Glacial Maximum – which saw the extinction of the megafauna and a dramatic decline in many other Australian mammals, like Tasmanian devils and thylacines – that reduced koala numbers most dramatically.

Inbreeding is thought to be bad because it increases the chances of parents sharing a deleterious gene, producing a disease or malformation in their offspring. But high levels of inbreeding can often reduce the frequency of these genes. When the unhealthy offspring either don't breed or don't survive, those genes are eliminated from the population. This is the principle behind many modern stock-breeding practices – leaving dairy cows and racehorses with negligible genetic variation but few problems with inbreeding.

While inbreeding depressions have been noted in several endangered populations, many species thrive with very little genetic diversity. Humans, for example, are a young, slow-breeding species that hasn't been around long enough to accumulate mutations. In effect, the entire world's population of 7.9 billion people appears to be descended from around 2000 people – the size of a small country town. That hasn't stopped us, however, from becoming one of the most widespread, abundant and physically variable terrestrial vertebrates on the planet. Genetic diversity might not be as important as simply lucking out in having the right genes.

Overall, koala genetic diversity isn't particularly low. It's similar to other outbred large populations, like feral pigs or white-tailed deer. Genetic studies can provide a great deal of fascinating information about koalas, but worrying about the impact of low genetic diversity in an unknown future is a risky distraction from the very real and immediate threats facing koala survival today.

Humans have been able to significantly expand our habitat range by virtue of our ability to change our environment to suit us – in effect, behavioural adaptations that do not require genetic change. Koalas, to a lesser degree, show similar behavioural flexibility, allowing them to populate a wide range of eucalypt forests, despite their apparent specialisation. Koalas might be individually specialised – but collectively, as a species, they are generalists.

The more immediate necessity for koalas, though, is their ability to adapt to disease. One of the biggest problems that koalas on the east coast currently face is chlamydia.

Chlamydia is best known as a sexually transmitted disease in humans, which can cause infertility and reproductive problems in women. The bacterial disease can also pass from mother to infant, causing eye ailments and pneumonia. It is usually treated with antibiotics. Chlamydia is also widespread in animals – like guinea pigs, sheep and cattle – making it one of the most widespread bacteria in the world. As a result, it has long been assumed that the disease was introduced into the otherwise isolated landmass of Australia from livestock, particularly sheep, on the First Fleet.

Over the last few years, closer genetic research has suggested that chlamydia bacteria actually comprise several different species that target different hosts. The varieties that infect humans are usually distinct from those common in livestock or birds. The ones affecting koalas are *Chlamydia pecorum* – also prevalent in other hosts such as cattle, sheep and goats – which cause diseases of the eyes, urinary and reproductive tracts. *Chlamydia pneumoniae*, which leads to significant respiratory diseases in humans, can also be caught by koalas.

Despite being one of the most thoroughly researched aspects of koala biology, the origins of chlamydia in koalas is an ongoing puzzle. Genetic studies by researchers at the University of the Sunshine Coast seem to suggest that the chlamydia found in sheep is the most likely source of the varieties in koalas, and that this infection has probably crossed from sheep to koalas repeatedly since colonisation. Chlamydia bacteria are commonly shed in faeces, causing infection to spread rapidly through flocks of ground-feeding sheep. The transmission route from exclusively ground-dwelling sheep to predominantly arboreal koalas may seem unlikely, but perhaps a koala, having inadvertently stepped on a sheep's droppings, must have transferred the bacteria from its hand to its mouth while eating. Recently, the first complete genome sequence of *Chlamydia pecorum* from koalas revealed a much higher level of diversity across populations, suggesting that while some chlamydia infections may come from cattle or sheep, others might be unique to koalas. Wherever it comes from, though, once caught, chlamydia spreads rapidly within koala populations primarily through sexual activity.

This disease is horrendous in koalas. The bacteria are highly infectious, causing conjunctivitis, eye inflammation and blindness, chronic urinary tract infections and incontinence, pneumonia and infertility in both males and females. Breeding is substantially reduced in populations infected with chlamydia. The disease relentlessly degenerates the body, leading to a slow, drawn-out and painful death.

'Chlamydial cystitis is awful,' says one of the koala rescuers in the northern rivers area of New South Wales, where the disease is rife. 'The bladder wall thickens, they can't store urine, it runs out and their constantly wet bottom becomes ulcerated.'

'Animals can literally cry when urinating, it hurts them that much,' Professor Peter Timms, a chlamydia researcher, said.

Treatment options are limited, particularly for animals at a late stage of the disease. Because the koala's liver is supercharged for removing toxins from gum leaves, it is also supremely efficient at removing medicines from the body. A single dose that would cure chlamydia in a human might need to be repeated every day for a month for a koala. Such a massive dose of antibiotics would, in any case, destroy the very gut biota the koala needs to survive. The options are not promising. And in most cases, the damage has already been done. Most infected animals are already infertile and those that are treated frequently become reinfected.

This problem of recurring infections has led some to suggest culling afflicted individuals and letting the population come back.

'About half the koalas across Australia are infected,' says David Wilson, a professor of infectious diseases at the Burnet Institute in Melbourne. 'In closed populations, the majority can be infected – sometimes up to 80 per cent.'

'They're transmitting chlamydia to each other and many of them can't be healed. These koalas are in a lot of pain and if they're out of the time-range of antibiotics being effective the humane thing to do is probably to euthanize them.'

It's a rational argument and, in theory, a disease-free koala population could begin to recover in five or ten years. But it's not a popular suggestion. And it's not certain that disease is the only cause of the koala's problems in these areas. What role is being played by land clearing, development, deaths by dogs or cars, logging or climate change? And not all infected koalas suffer the same severe symptoms. Perhaps there are other ways of effecting a cure.

The incidence of chlamydia varies in different koala populations. In Queensland and New South Wales, some areas have up to 80 per cent infection rates, while the average is about 50 per cent. Most infected animals here have severe symptoms. In Victoria and South Australia, the incidence and severity are lower, largely due to the disease-free koalas that were moved to French Island in the late 1890s. The absence of any chlamydia in this founding group is thought to be one of the main reasons the population boomed and expanded so rapidly. And indeed, this has happened wherever French Island koalas have been introduced into disease-free habitats – such as on Kangaroo Island. Keeping chlamydia out of koala populations has proven difficult, though. Chlamydial infections seem to be expanding through the population in the Mount Lofty Ranges, largely founded by disease-free Kangaroo Island animals. The effects of the disease here, however, seem to be less severe. Although 47 per cent of the population is infected, only 4 per cent show clinical signs of the disease.

In other places, though, the impact of chlamydia has been dire, as illustrated by the koalas released in the Gariwerd or Grampians National Park in Victoria. The sandstone escarpment of the Grampians rises in dramatic relief from the flat Victorian farmlands that surround it. Protected from development by its rugged geology, the Grampians is an isolated refuge for wildlife and native vegetation, as well as a popular camping and hiking destination for holiday-makers. Bushfires in 1938 were thought to have wiped out the Grampians' endemic koala population, so in 1957 some 600 chlamydia-free koalas were translocated from French Island to the Halls Gap area, where they thrived.

I remember camping in the Grampians with friends in the late 1980s. I think it was the first time I saw a koala on the ground. Coming out of my tent one night, my torch flashed across the white rump of a koala sauntering across the lawn between the trees. It was a highlight of the trip, and the koala's ready visibility certainly contributed substantially to the area's charms. But although I didn't realise it at the time, the Grampian koalas were already in some trouble.

In 1947, another small population of koalas had been brought to the area, confined to an island in the middle of an artificial lake. This population was from Phillip Island, and these koalas probably had chlamydia. Their numbers grew, albeit more slowly than the ones from French Island, and surplus animals were eventually moved off the island, where they mingled with the main disease-free population. By the early 1990s, breeding by koalas in the area was rare and 84 per cent of the population was infected. Bushfires in 2006 and 2013 also didn't help. The koalas became so rare that none were seen in the Grampians at all between 2014 and 2020.

It's a sobering reminder of just how quickly koala populations can rise and fall, and how flourishing or even overabundant ones can almost entirely vanish.

Vaccination has long promised protection from the world's worst diseases. From its early beginnings in variolation – practised in China, Africa and Turkey – to the rapid scientific advances of the eighteenth and nineteenth centuries, vaccines have achieved amazing results. Smallpox, polio, tetanus, rubella, measles, mumps, diphtheria and pertussis have been all but eradicated across many parts of the world.

And there is similar hope for koalas. As Australians were lining up to receive their vaccines for COVID-19, koalas were beginning trials of a new chlamydia vaccine, albeit on a much smaller scale.

Professor Peter Timms and his colleagues at the University of the Sunshine Coast began trials of a vaccine after ten years of research. A series of small trials seemed to prevent vaccinated koalas from catching chlamydia, as well as halting the progress of the disease in infected individuals. A much larger roll-out began in October 2021 through regional zoos and wildlife hospitals with plans to vaccinate as many as 1000 koalas. In parallel, the researchers are working to have the vaccine product registered, so that it can be available even more widely in the future.

'We know the vaccine is safe,' says Peter. 'We know that it can reduce infection levels.'

Modelling suggests that even with a low level of vaccination – 10 per cent per year of young female koalas – the disease burden in the koala population could be significantly reduced within five or six years.

Perhaps without so much chlamydia, without the infertility and blindness, the long and painful deaths, koalas would stand a fighting chance against all the other challenges they face in Australia's shrinking forests – the roads and cars and dogs and houses, the changing climate. But would it be enough?

20

Expansion and Retreat

The onlookers stand on the side of the riverbank, pointing and shouting excitedly. The river is wide by Australian standards, which only means that it is too deep to ford and too broad to bridge by a fallen tree. There's a noticeable current, which not all rivers have in this dry continent. And something is trying to cross it.

A small head bobs above the surface. It is not a dog – it almost looks like a child, but it is not struggling in the flow. It moves purposely towards the observers on the bank. It adjusts its course to counter the flow of the river, ensuring that it reaches the riverbank exactly where it intends.

'It's a koala!' one of the observers shouts, as the animal rises from the water onto the bank.

The koala stops at this noise and looks up, as if it had only just noticed these strangers standing in its path. It gives itself a shake, fluffing up its fur and shedding an outer layer of dampness. The watchers continue to hoot noisily.

'What's it doing?'

'Oh, you poor little thing!'

The koala's ears rotate once, twice and then, without hesitation, it turns and walks determinedly back into the water, launching itself into the current and across the river again.

For an animal that lives in trees and is famed for not needing to drink much, koalas are surprisingly at home in the water.

The dramatic changes to Australia's landscape, vegetation and animals over the last 2 million years – the shifting river systems, the aridity, the extinctions – all fall into sharp relief against the rapid devastation over the last 200 years. Colonisation, industrialisation and urbanisation have left the landscape almost unrecognisable. Less than a quarter of Australia's native vegetation remains more or less intact, and none of it without some impact. And the most affected area correlates almost perfectly with the mid-altitude forests of the east and south coasts favoured by koalas. A comparison of reconstructed pre-1750 vegetation maps with current ones reveals how the heartland of this area has been scraped bare by agricultural land clearance, leaving only residual traces of patchy green around the edges. Of all the vegetation, eucalypt woodlands have been reduced the most.

In the first 100 years of settlement, over a million square kilometres of the richest and most fertile forests were cleared for farming. This was half of the land suitable for intensive farming; the forests on rocky, less arable soils were usually left alone. By the beginning of the twentieth century, forestry practices had become more established, managing wood production and protecting water catchments near urban areas. From the 1960s, the conservation movement has grown in an effort to shield the remaining forests from commercial exploitation and development.

In the 1990s, bans on the further clearing of native vegetation should have protected the forests, but exemptions and loopholes for agriculture and forestry have seen vast tracts logged or replaced by pasture, particularly in Queensland, where over 2.4 million hectares of forest were either cleared or re-cleared between 2010 and 2018. New South Wales cleared 662,000 hectares over the same period, while the remaining states combined cleared a similar area. Over the last twenty years, the Queensland koala population has dropped by 40 per cent, and the New South Wales population by 33 per cent.

It's not just the overall figures that are important, though. The pattern matters, too.

I most commonly see koalas emerging from the patch of bush at the bottom of our gully. They march through the gate near a wood-lot of blue gums, and take a shortcut across the paddock to reach the dense row of trees planted along our boundary line. They never stay long there. I think they use this patch of trees as a transit point between the larger forests on either side. These narrow corridors of trees and vegetation are lifelines for their movement, dispersal and survival, joining up the increasingly fragmented forest landscapes. The further apart the forest remnants are, the harder and more dangerous it is for koalas and other animals to reach them.

And while all forests are important for biodiversity and ecological functioning (like carbon capture, nutrient cycling, and water and air purification) – not to mention their impact on climate – some support higher levels of biodiversity than others. Forests associated with water catchments – rivers, lakes, aquifers and springs – are particularly rich in biodiversity, making them especially important to koalas. And yet these forests suffer from a double whammy. Australia has an appalling record in not only protecting the health of its scarce native forests, but also in looking after one of its most precious resources, freshwater.

Water limits everything in Australia, from wildlife to agriculture to our urban centres. For the most part, it is only abundant around the edges of the continent, in the more sparsely populated northern tropics, down the coastline east of the Great Dividing Range and in the southern forests, particularly in Tasmania and some parts of Victoria. For the dry inland in the rest of the country, the only significant rivers are part of one system – the vast Murray–Darling, which sprawls across broad, flat, rarely flooding plains that cover much of inland Queensland, New South Wales and Victoria, before reaching the sea in South Australia. Dividing up this scarce asset equitably and sustainably in the drier areas of states and regions has proven to be an intractable problem, let alone simultaneously ensuring that the river retains enough water to support the forests and animals that depend on it for survival. This internecine dispute between the states has paralysed management of the river system. The immense expansion of irrigated crops in the 1970s, particularly in the headwaters of the river system, has exacerbated the impact of the regular cycle of droughts that characterise the Australian climate. Only 3 per cent of the Murray–Darling river system remains unmodified – most of it is severely degraded. The remnants of the river red gum forests that once stretched along its waterways and over the floodplains are no longer revived by periodic flooding. The floodplains have been cleared and the forests have retracted to a narrow, fractured strip of remnant trees, choked with weeds, along the water's edge. Stark grey skeletons of ancient forests rise from the shallow waters of huge dams and artificial lakes – bleak monuments to the forests that once thrived, sheltered and sustained the river ecology. Mass fish kills and toxic algal blooms have become increasingly common.

Such broadscale environmental changes have had a huge impact on Australian animals. Australia has one of the worst records for mammalian extinctions in the world. Since colonisation in 1788,

100 endemic Australian species have become extinct, a sharp upturn in the underlying pattern of extinction over the last 2 million years. Thirty-four Australian mammals have disappeared in the last 230 years – around the same number of mammal species lost in the rest of the world over the same time. The list of threatened and vulnerable species is growing. And while many extinct species had small or island ranges, abundance is no protection. Many of our lost mammals, like the pig-footed bandicoot or the crescent nail-tail wallaby, had extensive habitats.

While the ranges of so many species are contracting, the territory and population of one mammal continues to expand and intensify. Humans have multiplied at an exponential rate, and our global population of 7.9 billion is expected to double within 100 years. Even in Australia, where population growth is only sustained by migration, our urban centres, particularly along the east coast, are spreading, outstripping growth and planning projections. Urban development in southern Queensland, on the 'koala coast' surrounding Brisbane, is particularly intense. Homes for koalas and other wildlife are rapidly being replaced by ones for humans, which bring with them a host of extra menaces – dogs, cars, roads, gardens, fences and swimming pools.

So frequent are mortalities that koala hospitals have been established for the animals, largely by volunteers through fundraising. Most of the patients are victims of vehicle collisions and they're usually male, as their higher level of mobility causes them to often cross paths with cars. More female patients come through in the breeding season. Dog attacks are another common source of mortality.

It's not that the koalas can't live with these changes. Often they can: if there are enough trees, of the right kind, for them to live in, in linear parks that follow old creeklines; if enough trees are left in the paddocks for them; if there are places for them to cross roads safely; if new urban developments retain old eucalypts and maintain habitat corridors; if dogs are managed and confined; if rural and urban fences are constructed for wildlife safety instead of as traps to entangle, ensnare and obstruct; if swimming pools have slopes and steps for animals to exit; if we take the time and make an effort.

Koalas are remarkably resilient to habitat degradation – even better than other species – but they cannot cope with all the challenges the modern world has unleashed upon them.

My neighbours regularly send me emails about the koalas in our area.

'There were three koalas bellowing last night. One was quite close but I couldn't tell if it was Burt or not,' one reports.

It reminds me how common koalas are here, how lucky we are to have so many. But I worry about whether this is viable – about whether the forest fragments can support them all. Where I live, surrounded by parks, the forest feels endless, but I know it is not. The Mount Lofty Ranges in Adelaide – recognised as a hotspot for biodiversity – has lost 90 per cent of its original vegetation cover. The forest I live near is just an isolated pocket of what was once here, now disintegrated by farmland and housing.

We've had a koala boom in recent years. The Mount Lofty Ranges contain one of the densest populations of koalas in the country – up to fifteen animals per hectare in some patches, compared to parts of New South Wales or Queensland where there are as few as one koala per 14 hectares. Despite being in the driest state in the country, and in one often considered to have no native forests (those worthy of a forester's attention, anyway), koalas have found the perfect patch.

Mount Lofty peaks at just 727 metres and has an average rainfall of 700 millimetres. The current distribution of koalas in South Australia – on Fleurieu Peninsula, western Kangaroo Island and the state's south-west corner – overlay almost exactly with the highest rainfall areas of 600–1000 millimetres. They are spreading out over the drier Adelaide Plains along treed parklands, and pop up in isolated pockets along the River Murray and on the tip of Eyre Peninsula.

Even within the ranges, the density of koalas varies greatly. Around Cleland, one of the early sites of introduction, they are less abundant than in the Morialta or Belair national parks. Perhaps it is due to differences in microclimate – rainfall, exposure, heat – or the distribution of their preferred manna gums, or maybe it's the different soils, their fertility or the trace elements of the rocks that lie beneath them.

Whatever the reason, this abundance of koalas here is not because the habitat is especially good or the trees are better. Although manna gums are highly nutritious and the expansion of southern blue gum plantations have provided new pastures for koalas, the forests here are not extensive and most are lightly wooded. Many would hesitate to even call these woodlands 'forests'.

Rather than humans encroaching into koala territory, koalas here are moving into human habitats. The city parks are now home to significant populations, including breeding females. The perils of urban living remain but are not enough to outweigh the benefits. As the koala population continues to grow, management plans in the southern states are focusing not just on protecting koalas, but also their habitats. We know what happens when the number of koalas outstrips their food supply.

Culling is not an option and is not a policy of any government agency. But even euthanising sick or injured koalas that cannot be saved on humane grounds is contentious.

'I've been called a murderer,' says one of the rangers tasked with the care of native plants and animals across the local reserves. 'More than once. How can they say things like that?'

It's a tough judgement on someone who has devoted their life to saving Australian animals from extinction.

'We don't get the privilege of just looking after individual animals. Our job is to manage threats to all native species and ecosystems – whether that's introduced predators, weeds or overgrazing by native animals,' she says. 'We have to try to keep the balance right for everything. And we don't have anywhere near the resources to do that.'

In 2013, koalas in the Otway Ranges reached critical capacity in the limited patches of manna gums that comprise their favoured food. Starving koalas crowded onto dying trees, just as they had on the Quail and French islands. Locals were concerned about the trees, but even more so about the koalas.

'The whole cape smelt of dead koalas. It smelt like death,' one local said. 'There are hundreds of acres of dead trees.'

Over several years, hundreds of koalas were euthanised – sparking claims of 'secret culls'. Koala density peaked at nearly fourteen koalas per hectare. Bitter experience with these crowded forests has led environmental authorities to conclude that one koala per hectare is sustainable for both the animals and the forests. But in the absence of disease, these koala populations seem to double every three years. Their numbers just keep increasing. It's a pattern Kath Handasyde has seen repeated through many Victorian forests over the last forty years.

'I can't stand seeing them starve to death,' she says. 'The forest was gone. All the trees were dead. It was horrifying.'

Is the problem that there are too many koalas? Or that there isn't enough forest? We still don't know why some introduced island populations (like on St Bees Island, off Mackay in Queensland) seem to be quite sustainable and stable, while others become out of control. For the moment, contraception seems to be the best tool available for slowing population booms.

Managing over-abundant koala populations in the southern forests is made much more difficult by the international public perception that northern koalas are becoming extinct. Ignoring the booming populations in the south – and dismissing them as somehow of lesser value on the basis of superficial differences in genetics or 'race' – means we are only looking at half of the picture. By failing to develop a national plan that takes into account the true diversity of koalas across the country, we are condemning thousands of southern animals to death by starvation and thousands of northern animals to death by disease.

It's taken quite a bit longer for koalas, and their popularity, to spread around the globe. Not everyone was convinced they were good ambassadors for Australia. In 1983, the federal tourism minister poured scorn on the idea.

'The belief of Americans that they are a lovely, cuddly little bear is fairly well exploded when they get here and pick one of the rotten little things up,' John Brown declared. 'They find it's flea-ridden, it piddles on you, it stinks and it scratches.'

It sounds like Brown had a really bad encounter with an upset koala. Like most mammals, they will urinate and scratch if they are stressed or scared. Wild koalas might suffer from ticks and mites, but they rarely seem to have fleas and by and large they smell quite nice. Sadly, Brown's misguided beliefs about the koala have passed into popular culture more firmly than any of his actual tourism campaigns. The koala, however, survived the reputational slight, going on to become the Australian animal icon in the 1984 Los Angeles Olympics and gain international fame.

Despite their growing popularity, the opportunity for people overseas to see a koala in a zoo was almost zero. For many years, they were just too difficult to keep alive. Even with a ready supply of eucalypt trees, many sickened and died when far away from their native forests.

The San Diego Zoo continued to focus on their koala program, receiving more koalas from Australia in 1928, 1951 and 1959, including a joey only discovered after they had arrived. The first koala to be born overseas was reported in 1960, swiftly followed by eleven more joeys through the 1960s. Prospects for a growing population dwindled, however, with Teddy, the last male koala at San Diego, dying in 1976.

In a bid to revitalise the colony, the Australian government waived export restrictions and allowed six koalas to be sent from Brisbane's Lone Pine Sanctuary in the same year as a gift to celebrate the bicentenary of the United States. In order to ensure a sustainable food supply for their 'flagship' Australian animal, San Diego Zoo also established a plantation of koala feed trees.

San Diego Zoo continued their successful koala colony, supported by an expanding dedicated eucalypt plantation. Today the plantation covers some 15 acres – around 6000 trees made up of 30 different eucalypt species. The favoured species year-round for their residents are river red gums and forest red gums, with manna gum, grey gums, swamp mahogany and other species being favoured, or at least accepted, at variable times of year. By 1983, San Diego Zoo had become so successful at breeding their koalas that they started loaning some to other zoos, with shipments of fresh leaves often flown out daily from San Diego. Under United States wildlife laws, zoos must contribute a proportion of their funds to in situ conservation activities for species considered at risk in their native habitat. San Diego's lucrative koala loans program provided important funding for koala research in New South Wales and Queensland.

Despite their huge appeal, koalas remain expensive animals to keep overseas. There are only ten zoos in the United States and a similar number in Europe that have koalas. Aside from the early short-lived attempts to keep koalas in England and a small colony in Israel, most of these animals are from the San Diego population. The United Kingdom and the United States both have eucalypt plantations to support their Australian residents. In 2018, a new colony was started at Longleat Safari Park in the United Kingdom, from South Australian animals – the first population of southern koalas to be established in Europe.

The legislation first introduced in 1927 to prevent the hunting of koalas and the export of their pelts had largely prevented the export of live animals, except for the early gifts to the United States zoos. By 1982, all Australian native vertebrates were protected from being hunted, exported or kept as pets unless specifically exempted. Detailed conditions for the overseas transfer of koalas began to be developed, including requirements for extensive transportation, exhibit design and staff training, and the receiving institutions had to demonstrate that they could provide the 'substantial amount of fresh eucalypt leaves daily' that koalas need to survive.

In 1984 Australia sent six koalas to Japanese zoos. The Japanese zoos built custom-made enclosures and had grown eucalypt plantations to provide plenty of food for their fussy new Australian guests.

The koalas arrived and settled in, but not all of them thrived. The koalas proved incredibly expensive to feed, three times more than an elephant and seventeen times more than a lion. A local newspaper reported that the cost of feeding just one koala was more than the annual salary of the city's mayor. Despite gum trees growing well as street trees in many parts of Japan, the keepers complained that the trees kept falling over in windy conditions.

'We were always busy running around,' one of them said. Not being able to use chemical pesticides meant that insects had to be removed by hand. I can imagine unruly gum trees proving difficult in a manicured Japanese garden, but the idea of picking off insects one by one from hardy toxic eucalypts defies my imagination.

In the late 1980s, one of the koalas in a Japanese zoo died. Its stomach was full of leaves, but the animal was emaciated and thin – it had starved to death with a full belly. A keeper tragically blamed himself. Was it some kind of disease? Sustaining koalas in captivity was clearly more than just having a steady supply of eucalypts on hand. Part of the success of the San Diego koala program has been an established dietary regime of eucalypt species. Each of the koalas at San Diego requires 200–400 of the heavily coppiced trees to keep it supplied with a variety of fresh new shoots year-round. No wonder koalas are expensive to keep.

This specificity of diet – dictated by the nutritional composition of different leaves from different species at different times, and limited by learnt preferences and perhaps the ecological composition of the individual koala's microbiome – causes difficulties for more than just international travel. Rescued animals face similar problems, particularly when their home range, and the eucalypts they have adapted to, have been burnt by fire.

So, is there a way to help translocated koalas transition to new diets more effectively? The team at Cleland Wildlife Park seems to think so.

'There's been so much more research into the gut biome,' Cleland's vet, Ian Hough, tells me, 'and we now realise it's not just about having good guys in terms of bacteria, it's about having precisely the right kinds of good guys.'

'It's been a huge problem for koalas rescued from bushfires because the trees they fed on have gone. If they won't eat the trees we can provide, they'll die. But we've found that if we inoculate them with the faecal matter from koalas currently eating the local diet, they almost immediately start eating those leaves as well.'

'How does that work?' I ask. 'How can they tell that they are now able to digest the leaves that they couldn't yesterday?'

'I don't know,' replies Ian, 'but it's a game changer. For the last couple of years, we've been routinely inoculating all the new koalas that arrive here, before they show any signs of trouble. It's worked really well.'

The process is yet to be tested scientifically, but it suggests that koalas tend to only eat certain kinds of eucalypts because their gut biome is uniquely tailored to the trees they have grown up eating. Move them somewhere else – give them different trees, even of the same species – and the koalas simply can't digest their food.

In the wild, under natural conditions, this is never a problem for koalas. They don't tend to move far, and if they disperse into new areas, such as after a bushfire, the process is slow and gradual over generations.

The koalas are not particularly enthusiastic about this inoculation process. I watch one koala, a large, rescued male who is healthy and eating well but not maintaining his weight. He'll receive a daily dose of a faecal inoculation – or poo smoothie – over ten days to ensure his gut bacteria are able to handle the local food. As the keeper approaches, a low growl rises from the koala's belly and he turns his head away. He knows what is coming, but there is no escape. The keeper gently puts one hand on the squirming koala's neck and quickly squirts the dose in the side of his mouth.

I'm amazed a wild, well-armed animal can take his medicine with such good grace. The koala reacts just like anyone would to something with a nasty taste – with outright disgust. It screws up its face, scrapes its mouth with its paws and drags its tongue across the roof of its mouth with a lot of loud 'lip-smacking'. If ever you needed proof that koalas are both highly sensitive to taste and astonishingly good-natured, this is it.

Cleland has many koalas that were rescued from the Kangaroo Island bushfires. It's unlikely they will ever return. The fires that raged across the island almost completely destroyed the forests and plantations that supported these animals. It will take time for the trees to recover, and the plantations may never be replanted.

It makes me think about all the koalas that were not rescued. In the summer of 2019–2020, the largest and most disease-free populations of koalas in Australia, established as a refuge on Kangaroo Island for protection against imminent extinction, were all but wiped out. In an increasingly unpredictable climate, fire has become one of the koala's greatest threats.

21

Under Fire

In January 2021, a bushfire devastated our neighbourhood. It started in the late afternoon. Looking out the front door, I could see a vast plume of smoke rising over the gum trees that line our road to the south. The smoke billowed and tumbled up into great mountains, darkening into shades of violet and purple, their underbelly shot through with rippling gold and orange. It looked like a bomb had gone off.

'Is it coming towards us?' asked one of my daughters.

The clouds drifted eastward, at the bidding of a gentle westerly breeze. The plume of smoke rose, but it didn't come any closer.

'Not at the moment,' I replied. 'Unless the wind changes direction, we should be fine.'

I checked the Bureau of Meteorology website. There was a change coming, with winds shifting south.

The clouds of smoke got larger. We turned on the scanner to listen to what was happening. No-one seemed sure where the fire was. It sounded like there was more than one ignition point. Reports came in from different locations, different roads, increasingly centred around the Scott Creek Conservation Park immediately to our south.

I watched the smoke plume, waiting for the waterbombers to knock the fire back, like they usually do, and for the smoke to diminish. It did not. It began looming above the shed and the trees, reaching towards us.

'That's getting closer, isn't it?' my daughter asked.

'Yes,' I replied. 'The fire will be here soon. We need to get ready.'

Bushfires are part of country life in Australia, integral to our vegetation, ecology and rural landscapes except for pockets of the wettest rainforests and the driest inland deserts. Anywhere there is sufficient vegetation that gets dry enough, will, inevitably, burn.

Australia burns by nature and by accident. It has done so with increased regularity since the climate began to dry during the Holocene. Charcoal and ash layer Australian stratigraphy, like the iridium layer marking the meteor strike that helped wipe out the dinosaurs. The Holocene charcoal layers mark a mass extinction, too – the retreat of the forests, the drying up of the inland lakes and the gradual loss of the megafauna.

The koalas managed to survive this mass extinction, despite a precipitous crash in numbers. They survived the next one as well, started by the arrival of European ships and the mass clearing of their habitats and culminating in our best efforts to hunt them to extinction in the early 1900s. Their powers of recovery are extraordinary. But can they survive a third, so soon after the last one?

Industrialisation has filled our atmosphere with skyrocketing levels of carbon dioxide, methane and nitrous oxide, creating a gaseous blanket that is trapping warmth beneath it, raising the world's temperature and changing the climate. The consequences for Australia's koala forests are severe. They are set to retract their range even further, and their leaves are likely to become less digestible. The humid northern forests may simply become too hot for koalas and their feed trees to live in, while longer, hotter summers in the south will increase thermal stress, kidney disease and mortality. And that's before we consider the increasing impact of fires.

Koalas have few defences against flames or extreme heat. At the tops of tall trees, protected from embers by thick fur, they may have been able to survive milder ground fires. With enough time, they can retreat with other wildlife to cool, damp gullies, sheltering in burrows and dams. But in the fierce, increasingly frequent fires that completely burn out the last remaining patches of forests, their capacity to survive and recover is much reduced. They cannot run away, cannot fight back, cannot defend themselves from these threats. They are living the apocalypse of our own making.

I live in a bushfire zone, so we plan for fires constantly. It's shaped the way we've built our house, developed our garden, managed the vegetation, planned our holidays and scheduled our work and leisure activities. It's a constant threat that requires ongoing planning, preparation and practice. Pretty much anyone who lives within a kilometre of a large area of native vegetation in Australia is vulnerable to bushfire. Excluding those in the wettest temperate and tropical rainforest areas, this includes most people living on the outskirts of our major cities.

I've watched bushfires approaching my childhood home on the Eyre Peninsula, doused embers in the northern suburbs of Sydney, known friends who lost homes in the 1980s during Ash Wednesday in

the Adelaide Hills and worked with survivors during the worst bushfire death toll in Australian history, when 173 people died in Victoria
during Black Saturday in 2009. I've written a book about Australia's
bushfire history and how we have, or haven't, responded to it, but
nothing could really have prepared me for the last few fire seasons
and, increasingly, what seems to lie ahead.

After many delays and reschedules, the pandemic restrictions have
lifted enough for me to make a quick trip to Brisbane. It's an opportunity to meet with local researchers and koala experts, but also to
finally see some northern koalas. I arrive at the Lone Pine Sanctuary
just before it opens. Even this early, it's hot standing in the small queue
of families waiting to enter.

Lone Pine is Australia's oldest and largest koala sanctuary. It was
started in 1927 by Claude Reid, at the height of the last Queensland
hunting season, and is now home to over a hundred koalas. I think
they have must have every single one of them out on display. There
are koalas everywhere, but not one of them is moving in this heat.
They collapse over branches, arms and legs dangling, heads drooping as if they are melting. They look hot – and extremely relaxed.

I'm struck by the contrast in size and shape from southern koalas.
They don't just weigh less on average – they also have much thinner fur, which makes them look even smaller and changes the shape
of their heads and ears. I finally understand why early paintings of
Sydney koalas are so dissimilar from the southern ones I'm used to,
with rounded heads and often downward-pointing ears, rather than
square, boxy heads and big fluffy upright radars. But as I lean on the
fence and watch the koalas stretch and yawn, scratch and occasionally

bellow, I forget about superficial differences. Three mothers sit amicably in the fork of a tree, their joeys exploring the surrounding leaves, until one of them decides she prefers another's spot and sits on top of her to force her out.

Small or big, northern or southern, they are all still koalas.

The koalas don't like our increasingly hot, dry summers. Like the eucalypts they feed on, koalas don't moult seasonally. It's not something marsupials seem to do. The thick pelts of the southern koalas are a problem in the longer and hotter summer weather. And even with shorter, thinner coats and smaller bodies, northern animals struggle with the rising temperatures.

In hot weather, koalas leave the light, lacy shade of the eucalypts and head for trees with thicker, darker shade, like acacias, pines or introduced deciduous trees. Shade trees are just as important as food trees. Koalas seek out cooler branches to hug and rest on, using them as 'heat sinks'. They even retreat to the ground in search of cooler, shady patches, to creeklines and gullies as well as gardens and swimming pools. But in a heatwave, when temperatures stay above 40 degrees Celsius for over a week and barely drop at night, there is no escape. We all just have to sit and wait and swelter.

The Australian bushfire season started early in 2019 and did not let up. Fires flared across south-east Queensland throughout September and continued through the coming months, burning not just the dry sclerophyll forests and grasslands, but pockets of fire-sensitive rainforest that had never been burnt before.

Fires pockmarked the New South Wales coast and hinterland, from north to south, from July until March the following year – unprecedented in their duration and wide distribution. Volunteer brigades from the west joined overseas aid and helped protect assets and mitigate the worst dangers. Smouldering fronts circled towns and farms, sweeping in first from one direction then another, keeping residents in a constant state of fearful anticipation. Smoke filled the lungs of those in inner-city Canberra and Sydney, and overwhelmed residents in the leafy urban fringe with an impending sense of dread.

Flames roared through forests and parks, up steep escarpments, threatening the isolated canyons which had protected the last surviving population of ancient Wollemi pines from fire for millions of years. This time, the 'dinosaur trees' needed air support to save them. But little could be done for the other forest inhabitants.

By mid-November, the hot, dry conditions had worsened in the south and lightning storms swept across Victoria, starting 150 fires in a single day, with the sudden catastrophic ferocity characteristic of southern fire conditions – a pattern that was repeated in Tasmania and South Australia in December. Fires in Western Australia threatened a rare mainland population of quokkas.

By the end of the year, temperatures hovered around 40 degrees, peaking at 47 degrees. A large fire burnt through several townships to the north of us, while fires on Kangaroo Island, barely contained, cast a dirty grey pall over brilliant summer skies. We were on high alert, checking the weather and listening for fire calls.

The thundering hum of a waterbomber roared above us, heading south, flying low and fast at full throttle.

'It's going to Kangaroo Island,' my husband said, checking traffic on the radio scanner. 'There's been lightning strikes in the park again.

They'll be a bugger to put out.'

Kangaroo Island lies just a few miles off the tip of the Fleurieu Peninsula, a geological continuation of the hills and ranges that extend all the way from the northern Flinders Ranges to the southern Adelaide Hills. The end closest to the mainland is farmed, cleared and populated with small towns and clusters of coastal holiday shacks. But the isolated western half of the island is a wild and spectacular landscape of rugged coastal cliffs and caves, cloaked in dense heathland and inland forests, mostly contained within the Flinders Chase National Park. It's a haven for rare and endangered species as well as over 48,000 koalas, one of the healthiest populations in the country. A fire here would be catastrophic.

'Hope they can put it out quickly,' I commented anxiously.

But there were too many ignition points, in densely vegetated areas that were inaccessible. The fires grew, spread, rekindled and merged, creating overwhelming fire fronts. A massive smoke plume billowed from the island as 96 per cent of the Flinders Chase National Park was burnt to a nightmare landscape of blackened skeletons and bare earth.

Historically, we have measured the impact of a bushfire by human lives lost, by houses and properties damaged, by economic cost, by livestock destroyed, by area burnt. But for the first time that I can remember, the Black Summer fires elicited an unexpected and international concern for our wildlife. Up to 19 million hectares was burnt in over 15,000 fires across the 2019–2020 fire season, an area that scientists estimated was home to 143 million mammals, 2.46 billion reptiles, 181 million birds and 51 million amphibians. Over 300 threatened species lay in the path of the fires, and a third of these lost more than half of their habitat. Because of the fires, 119 Australian species needed immediate intervention to save them from extinction.

But it was koalas that were the face of this wildlife tragedy.

Over 60,000 koalas are thought to have died in the fires – 40,000 on Kangaroo Island alone, in addition to 11,000 in Victoria, 8000 in New South Wales and 900 in Queensland. In New South Wales, the fires burnt a quarter of the remaining natural koala habitat.

The internet was flooded with images of rescued koalas – wrapped in blankets and bandages, being carried to safety, wounded, scorched, lost, thirsty, homeless and helpless. They clung to blackened trees or to the ash-streaked 'yellows' of tired volunteer firefighters. They sat, bewildered, in rows of laundry baskets or took shelter inside houses and drank from buckets and proffered bottles. They became a poster child for environmental anxiety and activism.

Friends and acquaintances contacted me from America, the United Kingdom and Europe.

'What can we do to help?' they asked. 'How can we save the koalas? We can't let them go extinct.'

I tried to explain to my anxious friends that, while there are many animals in danger of extinction, koalas are not one of them just yet.

'But we need to help the koalas,' they said.

Money flowed in for the koalas by the truckload. In the aftermath of the Kangaroo Island bushfires, the local wildlife rescue park started a fundraising campaign, hoping to raise $15,000 to 'help save Kangaroo Island's koalas and wildlife'. They raised $2.3 million. A tiny two-person family business suddenly had to upscale into a massive rescue operation. The resources dramatically improved the survival rates of injured koalas, but also extended to other wildlife,

establishing replacement habitats and developing long-term strategies for koalas on the island.

Port Macquarie Koala Hospital appealed for $25,000 to help
'thirsty koalas devastated by recent fires' and ended up with $7.9 mil-
lion. That's a lot of money for drinking stations and a small hospital. As
I read through the hundreds of comments online from donors express-
ing their admiration for these creatures, I was struck by the diversity of
contributors: a retirement village in New Zealand, a cat café in Japan,
children who baked cakes in the UK or donated their birthday money
in the US. Millions flooded in after Barack Obama tweeted about the
Australian fires and their impact on wildlife. Australian wildlife and
animal welfare agencies received almost $200 million for their work
rescuing koalas as well as many other animals.

'Koalas are the million-dollar babies,' a colleague tells me. 'They
raise more funds than any other species in the world.'

There is never enough money to protect the environment – to
control the introduced plants and animals that do so much damage,
to monitor incursions by neighbours, stock or illegal development,
to control erosion and restore damage, to monitor species, plan
for fires, to do the research, to educate and inform, to implement
improvements, to understand what is left and how to prevent it from
disappearing. Everyone sees the pictures of the flames and the burnt
animals, but the gradual, everyday disappearance of the trees, the
encroaching houses, the erosion of habitat goes largely unnoticed.
We are far better at rescuing animals in a disaster than saving them
from the dangers we've put them in.

The extra money saves many animals that would otherwise have
died. In New South Wales, they used trained dogs to locate injured or
starving koalas. On Kangaroo Island, aerial thermal imaging found

many additional animals clinging to the bare trunks, even after ground crews had already searched for them.

Dealing with burnt animals after fires is a devastating operation for farmers and wildlife carers alike. Many have to be euthanised and even with treatment the death toll is high. Cattle, sheep and kangaroos can't survive with burnt feet or broken limbs. Survival for any animal depends on the depth of the burns, the level of infection and stress. And it's not just the burns you can see that cause trouble. A neighbour who survived the 2009 Black Saturday fires tells me about the hospital staff carefully checking his nostrils before they attended to his burnt hands.

'Yeah, apparently if your nose hairs are gone, you're stuffed,' he says dryly. 'Means your lungs are burnt.'

It reminds me that a friend's horses died nearby that day, in a cleared paddock that should have been safe, simply from exposure to the superheated air.

It's always a struggle to know whether it's kinder to euthanise an injured animal or rehabilitate it. In the end, it comes down to how much pain the animal will suffer, whether they will be able to return to the wild and what is needed to look after them. There are barely enough government resources to care for healthy wild animals, let alone injured ones. Donations and volunteers make up the shortfall.

Koala rehabilitation is a long process. It can take a month or two of constant care for a koala with mild burns to recover. And even after release, they will need monitoring and health checks. Extra resources make a difference. Survival rates for koalas burnt in fires was once estimated to be less than 40 per cent, even with treatment, but after the 2019–2020 Black Summer fires more than half of all koalas brought in for treatment on Kangaroo Island survived.

But it's only a temporary reprieve. All the care in the world can't replace a lost forest, whether paved over by a freeway, buried under housing or burnt in recurrent bushfires that don't give the trees a chance to recover. What happens next summer when the fire season starts again?

I have learnt from past bushfires to keep the forest at arm's length. Gum trees are just too flammable to live among. I still live in a fire zone, surrounded by a conservation reserve to the south and east, water catchment reserve to the west and remnant bush to the north. But the neighbouring properties are open farmland, and we keep any flammable native trees at least 100 metres from the house. It's an important buffer and makes all the difference to survival in a fire. It's a luxury that wildlife doesn't have.

In January 2021, I watched as dark angry plumes of smoke continued to billow over our southern boundary. A golden glow illuminated the rolling clouds from underneath. Over the intermittent sounds of planes and fire engines, I could hear the dull, distant roar of the fire coming out of the conservation reserve and heading up the gully, towards us.

A large flock of yellow-tailed black cockatoos swept overhead – thirty or forty of them, heading north. They made none of their usual long shrieks or cackling wails; no creaking cries echoed through empty skies. On this day, their metre-wide wings beat steadily and they flew by in determined silence, as if conserving energy for a long flight.

Soon after came cresting waves of small birds darting overhead – mixed flocks, too small and fast to identify, all buzzing with frenetic alarm. Some of them had probably been tagged by the local bird banders, who have tracked the progress of individual birds in our local conservation reserve for over thirty years. I wondered how many would survive the devastation of their bushland refuge. Moments later, just on dusk, mobs of kangaroos flowed down the driveway with swift, unswerving velocity, flying across the paddock, their eyes firmly focused on reaching the bush at the back of our block. At least they had a chance to escape the immediate peril of the flames.

I saw no sign of any of the smaller, slower terrestrial animals, the reptiles and mammals. The fire was moving too fast for them to escape. Some of them might have sheltered underground or along creeklines. Others could have retreated into the hollows of the largest of trees, surviving if the fire was not too fierce. But the koalas would have had nowhere to go, nowhere to hide and no capacity to run. They would have sat in their trees as the heat and the flames rose around them – at the top of their own funeral pyre, watching it burn below. There could not have been any escape for them.

The fire arrived around nine o'clock in the evening, lighting up the horizon with a livid orange glow. Neighbours battled the flames that surrounded them, saving their homes as the fire passed. Others had left early. Most of their well-prepared houses survived, but two did not. The night air cooled and damped down the fire's progress, light winds backing and eddying. As the fire left the forest and reached the top of the hill, its energy was already spent. It flared up, illuminating the row of trees and dense wattles along our side fence in a spectacular wall of flames, before finally dying out.

A week and a half after the fires, our garden was full of new birds, a haven of unburnt green in the midst of a barren black world. The trees were full of noisy calls, and from time to time the lawn came to life with tiny red-browed finches that peeked above the short cut grass as if it were a forest. At night, the paddocks were filled with kangaroos, steadily mowing the grass.

I didn't see any koalas after the fires. There were none to be seen from the roads as I walked between the blackened pillars of trees that stood in mute testimony to the violence of the flames. We drove around the bottom of the park, where an arsonist had lit seven or eight small fires that spread and joined into one vast blaze. These multiple ignition points made it difficult for the waterbombers to extinguish the fire, allowing it to build so fiercely so quickly. One side of the road was untouched, the other scorched bare and desolate. An area that had always been a lush green forest of mature trees, a prized and much-loved reserve for wildlife, was charred and empty – a blackened apocalyptic wasteland.

I had hoped it might have been a cooler fire – which travels fast through the undergrowth, only burning lighter fuels – but it wasn't. It was patchy but large sections of the park were burnt with maximum intensity, incinerating everything from the mulch on the ground to the leaves on the trees – a full canopy burn. It was hard to see how anything could have survived that.

I also asked my neighbours if they'd seen any koalas.

'The only one I have seen since the fire, I buried,' one neighbour said bluntly.

'We haven't seen Bradley since the fire,' another neighbour reported. 'He was always in the big old tree out the front, but no-one's seen him since.'

Two koalas were found dead near a house that was destroyed in the flames. They'd fallen from a tall burnt eucalypt nearby. Another neighbour had several large old trees around their house which regularly had koalas living in them. She told me that in the days after the fire, they picked up ten bodies on their property, some young, some big, and buried them.

'There are four gums all of which survived the fire that the koalas favoured,' she said. 'One big resident koala was here the day before the fire. I think they used to get in under the old fences where the kangaroos had pushed their way through. It never occurred to me that they don't climb over the fences.'

She realised the horrible truth when she found a small body right next to a fence line.

'Just a couple more metres and it would have had a safe paddock that didn't burn,' she said, distressed.

Some koalas did survive. One approached a house and accepted some gum leaves offered to it. A few injured koalas were discovered with burnt paws and noses, including a mother and joey. A dozen or so found their way to the wildlife carers who went into the affected areas straight after the fire. A quick retrieval and swift treatment is probably one of the most important factors in rescued animals surviving. Most of the koalas that the carers found were on the edge of the burnt area, or in small unburnt patches. In one solitary unburnt tree, they found six koalas clustered together. Perhaps the animals moved faster than I thought.

I don't know how many koalas were living in the park before the fire. I'm not sure anyone does. I ask Professor Chris Daniels, an expert in koalas, who has devoted years to promoting the study and conservation of koalas in South Australia as well as across Australia.

'Nobody knows how many koalas died in this fire, because nobody knows how many were living there in the first place,' he says. 'Something like 40 per cent of the South Australian koala population died in the last three fires.'

'How many do you think there were in Scott Creek Conservation Park?' I ask.

'You get six koalas to a tree in Belair National Park,' Chris says. 'The last survey we did found thirteen to sixteen animals per hectare.'

The fire scar covers 2700 hectares. Even if there was only one koala per hectare, that's still a lot of animals.

The trees have now started to recover, their trunks growing a thick coating of green shaggy leaves from epicormic buds under their bark. But it will be years before they can support a new population of koalas. Even so, a few are already taking advantage of the regrowth. Koala numbers seem to be increasing after the Black Summer fires on Kangaroo Island and in New South Wales, and they seem to be doing well on new shoots. Eight months after the fire, a friend finds two koalas making a meal of the lush regrowth on one of the most intensely burnt ridges in the middle of the nearby conservation park. They are, indeed, tough little creatures.

In the last twenty years, the fire conditions in Australia have dramatically worsened. We are getting more fires, more frequently over bigger areas and over a longer fire season. Areas of forest that have never burnt before and have no resilience to burning are now under threat. And the length of time between fires is now too short to allow them to recover. These shifts are being driven by the changes in our climate, particularly the increase in greenhouse gases. None of this is good for the koalas nor any other mammal. Increasingly, we are set to lose more and more of our forest areas, and the animals that live there.

Over the course of their evolutionary history, koalas have responded to climate change, disease, changing forests, increasing aridity, predation and hunting. And they have survived. But the Black Summer fires proved just how easy it is to lose a huge number of koalas in just a few months. It's a perfect storm for catastrophic extinction. No matter how resilient they are, I'm not sure how well they are going to cope with all threats they are currently facing, nor the speed with which they are intensifying.

And to be honest, I'm not entirely sure how well we are going to cope as a species either.

22

In a Perfect World

I am trying to imagine a perfect world for koalas – with perfect trees, perfect soil and a perfect climate. There is, of course, no such thing, but it's a bit like one of those simulation games my kids used to play where they'd build a zoo, starting with just one enclosure then expanding over time while working out how to keep each animal happy and healthy. We've changed our world so much that it's easy to forget what kind of world koalas evolved in, what they are adapted for and what they need. Our baselines shift and we see our damaged and denuded world as normal and it's hard to grasp just how much change we've wrought.

The east and south coasts of Australia would be covered in euca-lypt forests. No farms, no fences, no roads, no towns, no powerlines. Just trees as far as the eye can see. I try to picture what today's frag-mented, dissected, eroded and ravaged landscape would look like: an endless sea of grey-green spreading out over hills, valleys, moun-tains and plains from the ridges to the coast. Darker ribbons would thread like arteries across the landscape, demarcating the creeks and waterways, with their rich abundance of biodiversity and the largest, oldest trees of the koala's favourite species.

I imagine these ancient giants spreading out along the creeklines and through the catchments, irregularly yet perfectly positioned, tracing out a pattern of water and nutrients from the soil. An abundance of ideal trees occupying great swathes of forest – manna gums, blue gums, red gums, box and messmate.

The females and their young occupy the prime habitat, in the lower-altitude forests on nutrient-rich soils – not too warm and not too cold. Perhaps these females spread themselves out in irregular polygonal ranges, a bit over a hectare in size, just enough to support one koala creating, growing and feeding an infant or two from conception to maturity. Or perhaps the females' ranges overlap. Overlaying these neat neighbourly ranges are the larger ones of the males, encompassing four or five female ranges as well as encroaching on each other's. The males feed on different trees, minimising competition with the breeding females and their young and maximising their reproductive output.

Here, under these imaginary optimal conditions, the koalas would sit and sleep and eat and regularly breed – year in, year out – until eventually they grow old, their teeth wear down and they can no longer derive sustenance from the leaves. In this perfect world, the joeys would mature then move on to search for their own place in this material garden of Eden, finding their own home. Death brings periodic vacancies in the surrounding home ranges, or perhaps a new space might be carved out between two established larger ranges. The polygons wobble and rearrange until stability is reached.

If not, the dispersing young must find new areas. They spread out, over substandard habitat. Good enough to live on, but not good enough to raise a family. They radiate out, from a safe central range, further and further in search of other trees capable of supporting a koala for a while. If they are fortunate, they will find a new pocket, a patch unoccupied by koalas. If they are not, they may adapt to living on inferior soils, roaming across a larger territory than their mothers did but managing to scrape together enough nutrients and sustenance to raise young. The less optimal the habitat, the fewer young that can be produced, the harder it is to breed and the more vulnerable the population becomes. Some habitats might only be home to dispersing males who can survive on poorer forage – spartan single quarters rather than the plush suburban homes for families.

These forests cannot go on forever, though. Eventually you reach the sea, the mountains or the desert. The population cannot rise indefinitely. There is always a limit. So, the steady increase in koalas must always be matched by loss – death through old age, disease, predation or accident. Mortality and fecundity balance themselves on the fulcrum of food supply.

Age is the ultimate reaper, the inevitable degeneration of the bodies we occupy, but birth rates vary. It is easier to find mates in more densely populated areas. It is easier to successfully conceive and raise young in prime habitats with healthy well-fed females. In less productive, sub-optimal habitats, koalas breed much more slowly. In prime habitats, where koalas are more abundant, they breed much faster.

Trees are not just for habitat, though; they are also for protection. The more time koalas spend on the ground, the more at risk they are from ground predators, like dingoes and thylacines. The relationship between predators and their prey adds another balancing act. Predators rarely control the population of their major source of prey. Rather, it is the abundance of these common prey animals that limits the number of predators. If there are too many predators and they reduce the abundance of their main prey species, they too will decline.

But the relationship between rare prey items and predators is uncoupled. Foxes, for example, might be sustained for the most part by the abundance of rabbits. If the rabbit population crashes, so too does the number of foxes. But every now and again, foxes will also eat a rare bandicoot. They might even eat all the bandicoots. But it makes no difference to the fox population if the bandicoots are wiped out or not. They can still eat rabbits. For a more scarce species, like bandicoots and koalas, intermittent predation can have a big effect and is never a matter of balance.

Koalas are too rare to be a dominant food supply. They have no safety in numbers, so they must adopt other strategies – like retreating to the trees.

While koalas may have avoided their main ground-dwelling predators that once roamed Australia's forests, like dingoes and thylacines, humans were a different kind of predator altogether. The koalas had no defence against climbing humans, with hurled spears and clubs and nooses. Their only defence was hoping to avoid attention and making themselves scarce.

There are other reasons abundance is a disadvantage, other than depleting your food supply or attracting predators. As humans have vividly experienced through successive waves of epidemics and pandemics, dense and highly mobile populations spread diseases with deadly rapidity. Chlamydia and retrovirus may always have lurked within koala populations, kept in check by strong, healthy immune systems and social distancing. Disease may well have been a major evolutionary driver in pushing koalas apart, just as the need to breed pushes them together.

These forces push and pull at koalas – clustering them together around food and mates, pushing them apart in response to disease and over-grazing, increasing their numbers in response to food and social proximity, reducing them again in response to predators, disease or hunger. The scales swing from time to time before shifting back into balance, until suddenly there is a cataclysmic shock to the system. I see koalas differently now. They cling tenaciously to whatever space we leave for them, climbing our suburban fences and invading our gardens, stubbornly refusing to give ground in the face of the changes we have wrought.

It's a miracle they have held on for as long as they have.

In April 2012, the Australian federal government declared koalas to be 'vulnerable'. By early 2022, they had upgraded the koala to an 'endangered' species, signalling a rapid deterioration in its status.

Species can be endangered, or vulnerable to extinction, at a local, regional, state, national or international level. There is a big difference between an animal going extinct in an area where it had previously been common and it disappearing entirely from existence, although both are concerning. Koalas were listed as vulnerable at a state level in New South Wales and Queensland in 2012 and 2015, but are not listed as threatened in Victoria or South Australia despite having been considered 'extinct' there in the past. The current federal legislation of 'endangered' only applies to populations in Queensland, New South Wales and the Australian Capital Territory, which lies wholly within New South Wales.

Many regional populations of species are listed as threatened nationally – particularly on islands, subspecies or in ecologically isolated bioregions. But the koala is one of the very few species whose designation is defined by the administrative boundaries of mainland state borders. It's a decision driven largely by a desire to be seen to be doing something to protect a high-profile and immensely popular species that is declining in areas close to major cities, while avoiding the obvious anomaly of classifying an abundant and booming southern population as under threat.

Globally, the International Union for Conservation of Nature Red List of Threatened Species designated the koala in 2012 as 'vulnerable' because their population had declined by 29 per cent over the previous 18–24 years. This assessment recognised, though, that koalas were already close to being considered 'near threatened' and that climate change and other pressures were expected to magnify this decline over the next 20–30 years. The last ten years seems to have proven this prediction correct.

Classifying a species as vulnerable, threatened or endangered without addressing the underlying causes of the problem, however, does little more than generate a few headlines. It goes without saying that the best thing we can do for koalas, for the forests, for the ecosystems that sustain us – quite literally, for life as we know it – is to stabilise our climate as best we can. The agricultural food-production systems that now feed our enormous human population have evolved during a period of remarkable climate stability over the last 10,000 years. We disrupt that stability at our own peril.

Just as our own future depends on our farms, the survival of koalas depends on their forests. In the last 200 years, we have reduced the Australian forests to the smallest area they have ever occupied. The forests are fragmented and isolated – which has simultaneously led to overpopulation and population decline. Isolated pockets of habitat prevent koalas from dispersing. This creates dense populations of largely disease-free animals in favoured tree patches, while massively

increasing mortality from disease in the northern forests. These fragmented and sometimes degraded habitats also increase the impact
of predation in more open forests, particularly on breeding females,
while the growing pressure on our water systems substantially reduces
the availability of high-productivity forest needed to successfully raise
young joeys to maturity.

We need to continue to manage, research and protect koala populations in the wild through vaccination and health programs. We
would be wise to continue captive breeding and the establishment
of disease-free, genetically diverse populations as a safety measure.
Ultimately, though, habitat protection remains the key goal. If we
can't manage native vegetation sustainably – if we can't protect our
forests – then we cannot protect the koalas, let alone any of the other
species who are at an even greater risk of extinction.

Protecting, restoring, extending and connecting habitat is the
primary task. While the further clearance of remaining native vegetation is largely illegal in Australia, land clearance continues at pace,
particularly in Queensland and New South Wales. Agriculture, forestry and mining are all exempted in various ways from protective
requirements, making Australia one of the few developed nations
listed as a global deforestation hotspot. The last thirty years of efforts
to protect koalas have not failed because we don't know how to protect them, but because we have not been able to protect the already
stressed forests they rely on.

Nor is it really enough to stop the decline. In truth, we probably
need to reverse it. Koalas might be able to cling on in a few strong-
holds, but many other species will not. From orange-bellied parrots
to mountain pygmy possums, a great many species are likely to disappear in the coming years and decades.

Healthy koala populations depend on the remaining large old trees being genuinely and effectively protected, as well as on the next generation of trees being able to replenish and expand a network of healthy forest ecosystems. These trees provide the structural framework for many other species – insects, reptiles, mammals and birds. Koalas are not anywhere near the most endangered species in Australia, but they are certainly an effective flagship species for conservation, leading the way for a broader protection of our wildlife and a greater engagement with the natural environment we all need to care for.

I'm wondering what will happen when spring is in full flow next month and the koalas begin their long-distance calls to their mates for their annual rendezvous. Will they call, from the unburnt bush to our north or the edge of the burnt-out park? Will they call only to hear silence echoing over the burnt-out valleys to the south?

I have a tray of seedlings in my garden, waiting to be planted out. They are tube stock, grown in long thin pots that suit the fast-growing tap roots that Australian natives typically send straight down into the soil. Their survival depends on getting their feet into some damp soil deep underground.

I don't need to plant out trees. The burnt area will rapidly revegetate with a thick forest of 'fireweeds' – native wattles, gums and banksias, which open their seed pods in response to the smoke and heat to reseed on the clean, bare, burnt soil.

But planting trees makes me feel like I'm doing something to help. Assisting nature to recover helps us to recover too from the devastation that bushfires leave behind. I check the ground for seedlings and notice that during the night a hungry mob of kangaroos has not only mowed all the weedy grasses down to stubs, but also neatly snipped off every little seedling that has emerged in the burnt area. Two steps forward, one step back. Nature does not always work the way we want it to.

I order some tree guards, wire and stakes to put around the new seedlings. A friend has brought around some young gum trees she grew on her farm nearby. She had to move when her parents sold the property, and she misses the giant red gums that stretched over her dam and shaded her horses. When she noticed some red gum saplings coming up in a pot, she took them with her. But she doesn't have anywhere to plant them in her rented garden, and even at just a year old they are already bursting from their large pots. They need to go in the ground.

'Do you have somewhere for them?' she asks. 'They need a good home.'

I do. Down in the unburnt patch of bush, we have been slowly removing weeds from a creekline. Bracken ferns, reeds and wattles now shade the creekline that was once filled with weedy blackberries, gorse and broom. Some weedy plums and willows have infiltrated the head of the dam. We remove them to make a small clearing. It's the perfect spot for a red gum.

We dig into the rich alluvial soil, freeing the young tree from its pot. Perhaps I should just let nature decide what tree should grow here, but there is nothing pure about conservation any more. We all just have to do the best we can. So, we spend more time weeding the bush than we do our garden. We control the feral foxes and rabbits, delight in the orchids that emerge from the leaf litter, and cheer when we find traces of bandicoot digs in the soft soil nearby.

We have done so much damage to nature, we can't really expect it to fix itself. For better or worse, we are part of this ecosystem, and we all must do whatever we can to help it function better – even if that is simply growing plants on the balcony or offering refuge to wildlife. Whatever little you do, nature repays many times over.

The sapling looks like it belongs. Soon I will plant more red gums and manna gums as well. They will join the blue gums the koalas already visit and will hopefully one day provide food for future generations. Some of my neighbours are also planting their gullies and creeklines with koala food trees, as well as trees for other wildlife – hakeas for the black cockatoos, goodenias for the butterflies. And nest boxes for the hollow-nesting birds and mammals until the trees are old enough to provide natural ones. We're linking up corridors of prime riparian habitat for wildlife throughout the area: joining the parks with reserves with private land with roadsides, with gardens, with remnant pockets of bush, with treeways across farmland, all the way from the hills to the city. It's a plan for the future – for decades ahead.

Once the planting is done, I pack up the shovels and buckets and prepare to walk back up the hill to the house. One day – long after I am gone – I hope this tree will fill the gully with vast branches. That it will be home to an abundance of birds and mammals and insects and plants and fungi. That its hollows and crevices will be filled with life, and that high up in the canopy a resilient female koala, a young joey growing in her pouch, is safe in the sheltering, nurturing branches of her new home.

ACKNOWLEDGEMENTS

The story of the koala has taken me into the distant past, across continent and cultures and through an incredibly wide range of knowledge systems: botany, ecology, Indigenous knowledge, evolution, palaeontology, anatomy, conservation biology, history, toxicology, psychology, veterinary and nutritional science, and animal behaviour. I am indebted to the great many researchers, conservationists, field naturalists, veterinarians, historians, animal husbandry experts, koala enthusiasts and observers who have provided me with such a rich resource of material to work from.

I am particularly grateful for the guidance and advice of Chris Daniels, founder of Koala Life, as well as Ian Hough, research manager at Koala Life, and koala keepers Ashleigh Hunter and Cassie at Cleland Wildlife Park for sharing their knowledge and time so generously and introducing me to their koalas. I am also grateful to staff from the South Australian Department of Environment and Water, and ForestrySA for sharing their experience of managing koalas and their habitat. And a special thanks go to the local volunteers at 1300KOALAZ, SAVEM and the many other volunteer wildlife organisations around the country who take on such a big role caring for our wildlife, particularly in the dreadful aftermath of recent bushfires.

Many researchers have provided their time, support and expertise for this project. My sincere thanks to Corey Bradshaw, Karen Burke da Silva, Aaron Camens, the late Mike Corballis, Simon Cropper, Bill Ellis, Ian Gibbins, Kath Handasyde, David Lindenmayer, Ben Moore, Dean Nicolle, Gilbert Price, Gavin Prideaux, Rolf Schlagloth, David Stemmer, Jess Taubert, Peter Timms, Vera Weisbecker, Rod Wells and Karen Viggers for their assistance. I am grateful for their efforts to bring me up to speed in many complex and rapidly changing disciplines; however, I take responsibility for the final conclusions and interpretations drawn in this book, and any resulting errors or misinterpretations.

This koala journey took a very different path to the one I expected. The pandemic prevented me from visiting many of the interstate researchers, koala habitats and facilities, but opened up unexpected local opportunities. Thank you to the caving groups FUSSI and Cave Exploration Group of South Australia for letting me join in their trip to the Corra-Lynn cave, and to Graham Pilkington for leading the trip and sharing his extensive experiences. A regional writers' residency funded by Writers SA to Coober Pedy and the far north turned out to be perfect location for writing about koala fossils and appreciating the scale and impact of climate change. Thanks also to my neighbours, extended family, local and online community for providing me with regular koala stories, anecdotes, updates and photos. I'd particularly like to thank Lily McAndrews for volunteering as an enthusiastic US-based volunteer research assistant and Allayne Webster and Katrina Germein for their unquenchable enthusiasm for koala-spotting.

I am indebted to Sophy Williams, Rebecca Bauert and the team at Black Inc. for initiating and nurturing this book to fruition, as well as to Amy Cherry from W.W. Norton for taking it to a worldwide audience.

And last, but never least, I owe a great debt to my family – my husband and daughters – for supporting my writing journey and sharing my love of the Australian bush and its wildlife.

BIBLIOGRAPHY

There has been a vast amount of research done on koalas, particularly in the last twenty years. This bibliography is not intended to be comprehensive but provides general publications as well as some specific references used to support information provided in the text.

FURTHER READING

Jackson, S. (2007). *Koala: Origins of an Icon*. Allen and Unwin, Sydney.

Martin, R. and Handasyde, K. (1999). *The Koala: Natural History, Conservation and Management*. UNSW Press, Sydney.

Moyal, A. (2008). *Koala: A Historical Biography*. CSIRO Publishing, Canberra.

Phillips, B. (1990). *Koalas: The Little Australians We'd All Hate to Lose*. Australian National Parks and Wildlife Service, Canberra.

Serventy, V. and Serventy, C. (2002). *Koalas*. Reed New Holland, Frenchs Forest.

REFERENCES AND NOTES

I – Into the Woods

This reconstruction is based on an unpublished account of radiotracking data from Kangaroo Island in the 1950s, which revealed a pattern of exploration by introduced koalas from their original location in a patch of manna gums to new habitats. It took many years for the small founding population of koalas on the island to either find, or adapt to, the habitat.

1 – Koalas Rare and Plenty

Notes: The scientific names of all species mentioned in the text are included in the index.

Anon. (1937). Adelaide's Koala Farm. *News* (Adelaide), 30 August, 10, http://nla. gov.au/nla.news-article130971725.

Department for Environment and Water, South Australia. (2022). Managing

koala populations in South Australia, https://www.environment.sa.
gov.au/topics/plants-and-animals/living-with-wildlife/Koalas/koala-
conservation-and-management/managing-koala-populations-in-south-
australia.

Lunney, D., Urquhart, C.A. and Reed, P. (eds) (1988). *Koala Summit: Managing Koalas in New South Wales*. NSW National Parks and Wildlife Service, Sydney.

Troughton, E.L.G. (1938). Australian mammals: their past and future. *Journal of Mammalogy*, 19(4), 401–11.

Woinarski, J.C.Z., Braby, M.F., Burbidge, A.A., Coates, D., Garnett, S., Fensham, R.J. et al. (2019). Reading the black book: the number, timing, distribution and causes of listed extinctions in Australia. *Biological Conservation*, 239, 108261.

II – From Fossils and Bones

This reconstruction of the habitat on Yorke Peninsula in the Pleistocene/Pliocene is very broadly inspired by the faunal list in Pledge, N.S. (1992). The Curramulka local fauna: a new late Tertiary fossil assemblage from Yorke Peninsula, South Australia. *Beagle: Records of the Museums and Art Galleries of the Northern Territory*, 9, 115–42.

2 – Dropbears in the Family

Notes: A video of the dropbear incident described in this chapter can be found at 'Scottish reporter tricked into wearing protective gear for "drop bears"', *7News*, South Australia, 13 January 2020, https://www.youtube.com/watch?v=KCGUNpzjD6M.

Dick, S. (2020). Wombats now protected all over Victoria after outrage over hunting lodge. *The New Daily*, 6 February, https://thenewdaily.com.au/news/national/2020/02/06/wombat-killing-laws-fixed.

Home, E. (1808). An account of some peculiarities in the anatomical structure of the wombat, with observations on the female organs of generations. *Philosophical Transactions of the Royal Society of London*, 98(XIX), 304–12.

Hooper, C. (2018). A wombat, a koala and a rabbit in a burrow. *Pursuit*, https://pursuit.unimelb.edu.au/articles/a-wombat-a-koala-and-a-rabbit-in-a-burrow.

Jackson, S. (2007). *Koala: Origins of an Icon*. Allen and Unwin, Sydney.

Troughton, E.L.G. (1938). Australian mammals: their past and future. *Journal of Mammalogy*, 19(4), 401–11.

3 – The Lakes District

Notes: The full species list of fossil koala mentioned in this chapter includes *Madakoala devisi, Madakoala wellsi, Perikoala palankarinnica* and *Perikoala robustus*.

Ambrose, G.J., Callen R.A., Flint R.B. and Lange, R.T. (1979). Eucalyptus fruits in stratigraphic context in Australia. *Nature*, 280 (5721), 387–89.

Black, K.H., Price, G.J., Archer, M. and Hand, S.J. (2014). Bearing up well? Understanding the past, present and future of Australia's koalas. *Gondwana Research*, 25(3), 1186–201.

Mather, E.K., Lee, M.S.Y., Camens, A.B. and Worthy, T.H. (2021). An exceptional partial skeleton of a new basal raptor (Aves: Accipitridae) from the late Oligocene Namba formation, South Australia. *Historical Biology*, DOI: 10.1080/08912963.2021.1966777.

Meredith, R.W., Westerman, M. and Springer, M.S. (2009). A phylogeny of Diprotodontia (Marsupialia) based on sequences for five nuclear genes. *Molecular Phylogenetics and Evolution*, 51, 554–71.

Rich, T.H., Lawson, P.F., Vickers-Rich P. and Tedford R.H. (2019). R.A. Stirton: pioneer of Australian mammalian palaeontology. *Transactions of the Royal Society of South Australia*, 143(2), 244–82.

Simpson, G.G. (1926). Mesozoic Mammalia, IV; the multituberculates as living animals. *American Journal of Science*, 5(63), 228–50.

Worthy, T., Camens, A., Mather, E.K., Blokland, J.C. and Lee, M. (2021). Meet the prehistoric eagle that ruled Australian forests 25 million years ago. *The Conversation*, 28 September, https://theconversation.com/meet-the-pre-historic-eagle-that-ruled-australian-forests-25-million-years-ago-168249l.

4 – From the Gulf to the Sea

Notes: The full species list of fossil koalas mentioned in this chapter includes *Invictokoala monticola, Koobor notabilis, Koobor jimbarratti, Litokoala garyjohnstoni, Litokoala dicksmithii, Litokoala kutjamarpensis, Nimiokoala greystanesi, Phascolarctos stirtoni* and *Priscakoala lucyturnbulli*.

Arman, S.D. and Prideaux, G.J. (2016). Behaviour of the Pleistocene marsupial lion deduced from claw marks in a southwestern Australian cave. *Scientific Reports* 6(1), 21372.

Bartholomai, A. (1968). A new fossil koala from Queensland and a reassessment of the taxonomic position of the problematical species *Koalemus ingens* De Vis. *Memoirs of the Queensland Museum*, 15(2), 65–72.

Black, K.H., Camens, A.B., Archer, M. and Hand, S.J. (2012). Herds overhead: *Nimbadon lavarackorum* (Diprotodontidae), heavyweight marsupial herbivores in the Miocene forests of Australia. *PLOS One*, 7(11), e48213.

Black, K.H., Price, G.J., Archer, M. and Hand, S.J. (2014). Bearing up well? Understanding the past, present and future of Australia's koalas. *Gondwana Research*, 25(3), 1186–201.

Louys, J. and Price, G.J. (2015). The Chinchilla Local Fauna: an exceptionally rich and well-preserved Pliocene vertebrate assemblage from fluviatile deposits of south-eastern Queensland, Australia. *Acta Palaeontologica Polonica*, 60(3), 551–72.

Martin, H.A. (2006). Cenozoic climatic change and the development of the arid vegetation in Australia. *Journal of Arid Environments*, 66, 533–63.

McDowell, M.C., Prideaux, G.J., Walshe, K., Bertuch, F. and Jacobsen G.E. (2015). Re-evaluating the Late Quaternary fossil mammal assemblage of Seton Rockshelter, Kangaroo Island, South Australia, including the evidence for late-surviving megafauna. *Journal of Quaternary Science*, 30, 355–64.

Price, G.J. (2008). Is the modern koala (*Phascolarctos cinereus*) a derived dwarf of a Pleistocene giant? Implications for testing megafauna extinction hypotheses. *Quaternary Science Reviews*, 27(27–28), 2516–21.

Price, G.J. and Hocknull, S.A. (2011). *Invictokoala monticola* gen. et sp. nov. (Phascolarctidae, Marsupialia), a Pleistocene plesiomorphic koala holdover from Oligocene ancestors. *Journal of Systematic Palaeontology*, 9(2), 327–35.

Price, G.J., Zhao, J.X., Feng, Y.X. and Hocknull, S.A. (2009). New records of Plio-Pleistocene koalas from Australia: Palaeoecological and taxonomic implications. *Records of the Australian Museum*, 61(1), 39–48.

Woodburne, M.O., Tedford, R.H., Archer, M. and Pledge, N.S. (1983). *Madakoala*, a new genus and two species of Miocene koalas (Marsupialia: Phascolarctidae) from South Australia, and a new species of *Perikoala*. In M. Archer (ed.), *Possums and Opossums: Studies in Evolution*, vol. 1. Surrey Beatty and Sons/Royal Zoological Board of NSW, Sydney, 293–317.

5 – A Giant at the Foot of the World

Note: Fossil koalas mentioned in this chapter include *Phascolarctos yorkensis*.

Archer, M., Black, K. and Nettle, K. (1997). Giant ringtail possums (Marsupialia, Pseudocheiridae) and giant koalas (Phascolarctidae) from the late Cainozoic of Australia. *Proceedings of the Linnean Society of New South Wales*, 117, 3–16.

Grand, T.I. and Barboza F.S. (2001). Anatomy and development of the koala, *Phascolarctos cinereus*: an evolutionary perspective on the superfamily Vombatoidea. *Anatomy and Embryology* 203(3), 211–23.

Phillips, B. (1990). *Koalas: The Little Australians We'd All Hate to Lose*. Australian National Parks and Wildlife Service, Canberra.

Pilkington, G. (1985). Corra-Lynn 5Y1 9–10 March 1985. *Cave Exploration Group South Australia Newsletter*, 29, (4 April), 60.

Pledge, N.S. (1992). The Curramulka local fauna: a new late Tertiary fossil assemblage from Yorke Peninsula, South Australia. *Beagle: Records of the Museums and Art Galleries of the Northern Territory*, 9, 115–42.

III – Life in the Forest

This reconstruction is loosely based on a Victorian mountain ash forest.

6 – Anatomy of a Climber

Notes: Videos of different koala locomotion can be seen at Echidna Walkabout Nature Tours (2019) 'Koala climbing up a tree by bounding', https://youtu.be/evX4YNXQzL4 and Cohen, A. (2008). 'Adelaide: Koala running in the park', https://youtu.be/La0LgRmdTcY.

Clode, D. (2006). 'The Ape Case'. In *Continent of Curiosities: A Journey Through Australian Natural History*, Cambridge University Press, Melbourne.

Cutts, J.H. and Krause, W.J. (1983). Structure of the paws in *Didelphis virginiana*. *Anatomischer Anzeiger*, 154(4), 329–35.

Grand, T.I. and Barboza F.S. (2001). Anatomy and development of the koala, *Phascolarctos cinereus*: an evolutionary perspective on the superfamily Vombatoidea. *Anatomy and Embryology*, 203(3), 211–23.

Haskell, D.G. (2017). 'Interlude Maple'. In *The Songs of Trees*, Black Inc., Melbourne, 155–66.

Hennenberg, M., Lambert, K.M. and Leigh, C.M. (1997). Fingerprint homoplasy: koalas and humans. *Natural Science*, 1(4).

Nagy, K.A. and Martin. R.W. (1985). Field metabolic-rate, water flux, food-consumption and time budget of koalas, *Phascolarctos cinereus* (Marsupialia, Phascolarctidae) in Victoria. *Australian Journal of Zoology*, 33, 655–65.

Prevost, A., Scheibert, J. and Debrégeas, G. (2009). Effect of fingerprints orientation on skin vibrations during tactile exploration of textured surfaces. *Communicative and Integrative Biology*, 2(5), 422–24.

Vogelnest, L. and Allan, G. (2015). *Radiology of Australian Mammals*. CSIRO Publishing, Clayton South, Victoria, 98–105.

Warman, P.H. and Ennos, A.R. (2009). Fingerprints are unlikely to increase the friction of primate fingerpads. *Journal of Experimental Biology*, 212, 2016–22.

Weisbecker, V. and Nilsson, M. (2008). Integration, heterochrony, and adaptation in pedal digits of syndactylous marsupials. *BMC Evolutionary Biology*, 8(1), 160.

Yum, S.M., Baek, I.K., Hong, D., Kim, J., Jung, K., Kim, S. et al. (2020). Fingerprint ridges allow primates to regulate grip. *Proceedings of the National Academy of Sciences*, 117(50), 31665–73.

7 – The Eucalypt Empire

Anon. (1999). Functions of phosphorus in crops. *Better Crops*, 83(1), 6–7.

Byrne, M., Steane, D.A., Joseph, L., Yeates, D.K., Jordan, G.J., Crayn, D. et al. (2011). Decline of a biome: evolution, contraction, fragmentation, extinction and invasion of the Australian mesic zone biota. *Journal of Biogeography*, 38(9), 1635–56.

Clode, D. (2021). Paddock of Dreams. *Cosmos,* Summer 93, 43–56.

Gleadow, R.M., Haburjak, J., Dunn, J.E., Conn, M.E. and Conn, E.E. (2008). Frequency and distribution of cyanogenic glycosides in *Eucalyptus* L'Hérit. *Phytochemistry*, 69(9), 1870–74.

Jensen, L.M., Wallis, I.R., Marsh, K.J., Moore, B.D., Wiggins, N.L. and Foley, W.J. (2014). Four species of arboreal folivore show differential tolerance to a secondary metabolite. *Oecologia*, 176(1), 251–58.

Martin, R. and Handasyde, K. (1999). *The Koala: Natural History, Conservation and Management*. UNSW Press, Sydney.

Moore, B.D., Foley, W.J., Wallis, I.R., Cowling, A. and Handasyde, K.A. (2005). Eucalyptus foliar chemistry explains selective feeding by koalas. *Biology Letters*, 1(1), 64–67.

Noble, I. (1989). Ecological traits of the *Eucalyptus* L'Hérit. Subgenera Monocalyptus and Symphyomyrtus. *Australian Journal of Botany*, 37(3), 207.

Phillips, B. (1990). *Koalas: The Little Australians We'd All Hate to Lose.* Australian National Parks and Wildlife Service, Canberra.

Ungar, P.S. (2015). Mammalian dental function and wear: A review. *Biosurface and Biotribology*, 1, 25–41.

8 – You Are What You Eat

Cork, S. (1995). Life in a salad bowl. *Nature Australia*, Spring, 32–37.

Ellis, W.A.H., Melzer, A., Carrick, F.N. and Hasegawa, M. (2002). Tree use, diet and home range of the koala (*Phascolarctos cinereus*) at Blair Athol, central Queensland. *Wildlife Research*, 29(3), 303–11.

Jensen, L.M., Wallis, I.R., Marsh, K.J., Moore, B.D., Wiggins, N.L. and Foley, W.J. (2014). Four species of arboreal folivore show differential tolerance to a secondary metabolite. *Oecologia*, 176(1), 251–58.

Logan, M. (2001). Evidence for the occurrence of rumination-like behaviour, or merycism, in the koala (*Phascolarctos cinereus*, Goldfuss). *Journal of Zoology*, 255(1), 83–87.

Maged, E.S.M. (2016). The interaction between the gall wasp *Leptocybe invasa* and *Eucalyptus camaldulensis* leaves: A study of phyto-volatile metabolites. *Journal of Pharmacognosy and Phytotherapy*, 8(4), 90–98.

Martin, R. and Handasyde, K. (1999). *The Koala: Natural History, Conservation and Management.* UNSW Press, Sydney.

Moore, B.D., Lawler, I.R., Wallis, I.R., Beale, C.M. and Foley, W.J. (2010). Palatability mapping: a koala's eye view of spatial variation in habitat quality. *Ecology*, 91(11), 3165–76.

Teaford, M.F. and Walker, A. (1983). Dental microwear in adult and still-born guinea pigs (*Cavia porcellus*). *Archives of Oral Biology*, 28, 1077–81.

Troncoso, C., Becerra, J., Bittner, M., Perez, C., Sáez, K., Sánchez-Olate, M. and Ríos, D. (2011). Chemical defense responses in *Eucalyptus globulus* (Labill) plants. *Journal of the Chilean Chemical Society*, 56(3), 768–70.

9 – The Guts of the Problem

Notes: The koala has the second-largest caecum of any mammal (185 millimetres) irrespective of body size, after the Madagascan aye-aye *Daubentonia madagascariensis* at 656 millimetres long. Relative to body mass, the aye-aye tops the list with a ratio of 46, followed by the sugar glider at 15.7 and koala at 9.8. A video of koala bellowing can be found at Revkin, A. (2015). Lots of sound, no fury: Male koala bellows and mellows, https://www.youtube.com/watch?v=9fmvf3FOC4o.

Beal, A. M. (1990). Composition of sublingual saliva of the koala (*Phascolarctos cinereus*). *Comparative Biochemistry and Physiology Part A – Physiology*, 97(2), 185–88.

Blanshard, W. and Bodley, K. (2008). 'Koalas'. In Vogelnest, L. and Woods, R. (eds) *Medicine of Australian Mammals*, CSIRO, Canberra.

Cork, S. (1995). Life in a salad bowl. *Nature Australia*, Spring, 32–37.

Degabriele, R. (1981). A relative shortage of nitrogenous food in the ecology of the koala (*Phascolarctos cinereus*). *Austral Ecology*, 6, 139–41.

Forbes, W.A. (1881). On some points of the anatomy of the koala (*Phascolarctos cinerus*). *Proceedings of the Zoological Society of London*, 18 January, 180–95.

Johnson, R.N., O'Meally, D., Chen, Z., Etherington, G.J., Ho, S.Y.W., Nash, W.J. et al. (2018). Adaptation and conservation insights from the koala genome. *Nature Genetics*, 50(8), 1102–11.

King, C.H., Desai, H., Sylvetsky, A.C., Lotempio, J., Ayanyan, S., Carrie, J. et al. (2019). Baseline human gut microbiota profile in healthy people and standard reporting template. *PLOS One*, 14(9), e0206484.

Lee, A.K. and Carrick, F.N. (1989). *Phascolarctidae.Fauna of Australia 1b: Mammalia, Volume 31*. Australian Government Printing Service, Canberra.

Milton, G.W., Hingson, D.J. and George, E.P. (1968). The secretory capacity of the stomach of the wombat (*Vombatus hirsutus*) and the cardiogastric gland. *Proceedings of the Linnean Society of New South Wales*, 93, 60–68, https://www.biodiversitylibrary.org/part/47914.

Moore, B.D. and Foley, W.J. (2000). A review of feeding and diet selection in koalas (*Phascolarctos cinereus*). *Australian Journal of Zoology*, 48(3), 317–33.

Smith, H.F., Parker, W., Kotzé, S.H. and Laurin, M. (2013). Multiple independent appearances of the cecal appendix in mammalian evolution and an investigation of related ecological and anatomical factors. *Comptes Rendus Palevol*, 12(6), 339–54.

Ziółkowska, N., Lewczuk, B., Petryński, W., Palkowska, K., Prusik, M., Targońska, K. et al. (2014). Light and electron microscopy of the European Beaver (*Castor fiber*) stomach reveal unique morphological features with possible general biological significance. *PLOS One*, 9(4), e94590.

IV – A Life in Reflection

This reconstruction is based on descriptions of female koala behaviour during the mating season by keepers and ecologists.

10 – From Pouch to Piggyback

Notes: A video of poop feeding can be found at I Contain Multitudes: Microbe Minute, (2017). How eating poop is the key to koala's survival, https://www.youtube.com/watch?v=k4RQX1ci5TY.

Bercovitch, F.B., Tobey, J.R., Andrus, C.H. and Doyle, L. (2006). Mating patterns and reproductive success in captive koalas (*Phascolarctos cinereus*). *Journal of Zoology*, 270(3), 512–16.

Bobek, G. and Deane, E.M. (2002). Possible antimicrobial compounds from the pouch of the koala. *Phascolarctos cinereus. Letters in Peptide Science*, 8(3–5), 133–37.

Charlton, B.D. (2015). The acoustic structure and information content of female koala vocal signals. *PLOS One*, 10(10), e0138670.

Charlton, B.D., Ellis, W.A.H., Brumm, J., Nilsson, K. and Fitch, W.T. (2012). Female koalas prefer bellows in which lower formants indicate larger males. *Animal Behaviour*, 84(6), 1565–71.

Charlton, B.D., Ellis, W.A.H., McKinnon, A.J., Cowin, G.J., Brumm, J., Nilsson, K. and Fitch, W.T. (2011). Cues to body size in the formant spacing of male koala (*Phascolarctos cinereus*) bellows: honesty in an exaggerated trait. *Journal of Experimental Biology*, 214(20), 3414–22.

Charlton, B.D., Whisson, D.A. and Reby, D. (2013). Free-ranging male koalas use size-related variation in formant frequencies to assess rival males. *PLOS One*, 8(7), e70279.

Ellis, W., Fitzgibbon, S., Pye, G., Whipple, B., Barth, B., Johnston, S., Seddon, J., Melzer, A., Higgins, D. and Bercovitch, F. (2015). The role of bioacoustic signals in koala sexual selection: insights from seasonal patterns of associations revealed with GPS-proximity units. *PLOS One*, 10(7), e0130657.

Farley, K. (1993). Learning and teaching science: Women making careers 1890–1920. In Farley, K. (ed.) *On the Edge of Discovery: Australian Women in Science*, Text Publishing, Melbourne.

Gibbs, M. (1947). *The Complete Adventures of Snugglepot and Cuddlepie*, Angus and Robertson, Sydney.

Johnston, S., McGowan, M., O'Callaghan, P., Cox, R. and Nicolson, V. (2000). Studies of the oestrous cycle, oestrus and pregnancy in the koala (*Phascolarctos cinereus*). *Journal of Reproduction and Fertility*, 120, 49–57.

Krockenberger, A.K., Hume, I.D. and Cork, S.J. (1998). Production of milk and nutrition of the dependent young of free-ranging koalas (*Phascolarctos cinereus*). *Physiological Zoology*, 71(1), 45–56.

Larkins, D. (2017). Naughty natives: the secret love lives of Australian animals. *ABC News*, 14 February 2017, https://www.abc.net.au/news/2017-02-14/naughty-natives-love-life-of-australian-animals/8265026.

Martin, R. and Handasyde, K. (1999). *The Koala: Natural History, Conservation and Management*. UNSW Press, Sydney.

Minchin, K. (1937). Notes on the weaning of a young koala (*Phascolarctos cinereus*). *Records of the South Australian Museum*, 6, 1–3.

Shao, Y., Forster, S.C., Tsaliki, E., Vervier, K., Strong, A., Simpson, N. et al. (2019). Stunted microbiota and opportunistic pathogen colonization in caesarean-section birth. *Nature*, 574, 117–21.

Watchorn, D.J. and Whisson, D.A. (2020). Quantifying the interactions between koalas in a high-density population during the breeding period. *Australian Mammalogy*, 42(1), 28–37.

Whisson, D.A., Dixon, V., Taylor, M.L. and Melzer, A. (2016). Failure to respond to food resource decline has catastrophic consequences for koalas in a high-density population in southern Australia. *PLOS One*, 11(1), 12.

11 – Sociable Loners
Notes: The entertaining video of the mother and young koala can be found on The Koala Channel, https://www.youtube.com/watch?v=TD37JJs_Twk. The first known painting of a Powerful Owl by H.C. Richter (in Gould, J. (1840), *The Birds of Australia*, R. and J.E. Taylor, London) shows it holding a young koala as its prey. The quote about wombats hating everyone is from Wahlquist (2016).

Anon. (2021). Koala at Japanese zoo becomes the world's oldest in captivity, *The Japan Times*, 2 March, https://www.japantimes.co.jp/news/2021/03/02/national/oldest-koala.

Davies, N., Gramotnev, G., Seabrook, L., Bradley, A., Baxter, G., Rhodes, J., Lunney, D. and McAlpine, C. (2013). Movement patterns of an arboreal marsupial at the edge of its range: a case study of the koala. *Movement Ecology*, 1(1), 8.

De Satjé, O. (1901). *Pages from the Journal of a Queensland Squatter*. Hurst and Blackett, London, 307.

Ellis, W.A.H., Melzer, A. and Bercovitch, F.B. (2009). Spatiotemporal dynamics of habitat use by koalas: the checkerboard model. *Behavioral Ecology and Sociobiology*, 63(8), 1181–88.

Hambling, B. and Pavey C. (2008). Predation on koalas by breeding powerful owls. *Australian Field Ornithology*, 25, 140–44.

Le Soeuf, D. (1918). Food of the diurnal birds of prey. *Emu*, 18, 88–95.

Martin, R. and Handasyde, K. (1999). *The Koala: Natural History, Conservation and Management*. UNSW Press, Sydney.

Melzer, A., Tucker, G., Hodgson, J. and Elliott, B. (2003). A note on predation on koalas *Phascolarctos cinereus* by raptors, including wedge-tailed eagles *Aquila audax*, in Queensland. *Queensland Naturalist*, 41, 38–41.

Mitchell, P. (1990). Social behaviour and communication of koalas. In Lee A.K., Handasyde K.A. and Sanson, G.D. (eds), *Biology of the Koala*, Surrey Beatty, UK, 151–70.

Norman, J.A., Phillips, S.S., Blackmore, C.J., Goldingay, R. and Christidis, L. (2019). Integrating measures of long-distance dispersal into vertebrate conservation planning: scaling relationships and parentage-based dispersal analysis in the koala. *Conservation Genetics*, 20, 1163–74.

Robbins, A., Loader, J., De Villiers, D., Beyer, H.L. and Hanger, J. (2019). Predation by carpet pythons (*Morelia spilota*) is an important cause of mortality in a free-living koala (*Phascolarctos cinereus*) population in south-east Queensland. *Australian Veterinary Journal*, 97(9), 351–56.

Semmler, E. (2021). Central Queensland photographer turns koalas into talk of the town, ABC Capricornia, *ABC News*, https://www.abc.net.au/news/2021-11-17/koala-sightings-central-qld-emerald-good-news/100624968.

Wahlquist, C. (2016). Woman attacked by wombat thought she was going to die. *The Guardian*, 22 August, https://www.theguardian.com/australia-news/2016/aug/22/woman-attacked-by-wombat-thought-she-was-going-to-die.

Waser, P. and Jones, W. (1983). Natal philopatry among solitary mammals. *The Quarterly Review of Biology*, 58(3), 355–90.

Watchorn, D.J. and Whisson, D.A. (2020). Quantifying the interactions between koalas in a high-density population during the breeding period. *Australian Mammalogy*, 42(1), 28–37.

12 – When It's Smart to Be Slow

Notes: The video of the stranded koala being rescued can be found on *Inside Edition*, 2017, 'Stranded koala rescued by students in canoe', 31 August, https://youtu.be/R-j3bMCEoJ4. Examples of blogs about koalas not being very smart and having small brains include https://blogs.unimelb.edu.au/sciencecommunication/2017/09/17/koalas-not-the-smartest-tool-in-the-shed/ and https://koalainfo.com/koalas-have-unusually-smaller-brain. Videos of a koala in someone's car can be found at *USA Today* (2019) https://www.youtube.com/watch?v=9pmFSn9HORc.

Clode, D. (2011). *Killers in Eden*. Museum Victoria Publishing, Melbourne.

Cunningham, M. (2017). Koala takes canoe trip to safety after found stranded in the Murray River, *The Age*, 28 August, https://www.theage.com.au/national/victoria/koala-takes-canoe-trip-to-safety-after-found-stranded-in-murray-river-20170828-gy618a.html.

De Miguel, C. and Henneberg, M. (1997). Encephalization of the koala, *Phascolarctos cinereus. Australian Mammalogy*, 20, 315–20.

Grand, T.I. and Barboza., F.S. (2001). Anatomy and development of the koala, *Phascolarctos cinereus*: an evolutionary perspective on the superfamily Vombatoidea. *Anatomy and Embryology*, 203, 211–23.

Herzing, D.L. (2014). Profiling nonhuman intelligence: an exercise in developing unbiased tools for describing other 'types' of intelligence on earth. *Acta Astronautica*, 94(2), 676–80.

Taylor, J., Brown, G., De Miguel, C., Henneberg, M. and Rühli, F.J. (2006). MR imaging of brain morphology, vascularisation and encephalization in the koala. *Australian Mammalogy*, 28, 243–47.

Weisbecker, V., Blomberg, S., Goldizen, A.W., Brown, M. and Fisher, D. (2015). The evolution of relative brain size in marsupials is energetically constrained but not driven by behavioral complexity. *Brain, Behavior and Evolution*, 85, 125–35.

Weisbecker, V., Rowe, T., Wroe, S., Macrini, T.E., Garland, K.L S., Travouillon, K.J. et al. (2021). Global elongation and high shape flexibility as an evolutionary hypothesis of accommodating mammalian brains into skulls. *Evolution*, 75, 625–40.

13 – Sensory Overload

Amador, G.J., Mao, W., Demercurio, P., Montero, C., Clewis, J., Alexeev, A. and

Hu, D.L. (2015). Eyelashes divert airflow to protect the eye. *Journal of the Royal Society Interface,* 12(105), 20141294.

Arrese, C.A., Hart, N.S., Thomas, N., Beazley, L.D. and Shand, J. (2002). Trichromacy in Australian marsupials. *Current Biology,* 12, 657–80.

Arrese. C.A., Oddy, A.Y., Runham, P.B., Hart, N.S., Shand, J., Hunt, D.M. and Beazley, L.D. (2005). Cone topography and spectral sensitivity in two potentially trichromatic marsupials, the quokka (*Setonix brachyurus*) and quenda (*Isoodon obesulus*). *Proceedings of the Royal Society B,* 272, 791–96.

Brischoux, F., Pizzatto L. and Shine R. (2010). Insights into the adaptive significance of vertical pupil shape in snakes. *Journal of Evolutionary Biology.* 23(9), 1878–85.

Charlton, B.D. (2014). Discrimination of sex and reproductive state in koalas, *Phascolarctos cinereus,* using chemical cues in urine. *Animal Behaviour,* 91, 119–25.

Charlton, B.D. (2015). Chemosensory discrimination of identity and familiarity in koalas. *Behavioural Processes,* 119, 38–43.

Grant, R.A. and Goss, V.G.A. (2021). What can whiskers tell us about mammalian evolution, behaviour, and ecology? *Mammal Review,* 52(1), 148–63.

Harlow H.F. (1958). The nature of love. *American Psychologist,* 13(12), 673–85.

Hemsley, S., Palmer, H., Canfield, R.B., Stewart, M.E.B., Krockenberger, M.B. and Malik, R. (2013). Computed tomographic anatomy of the nasal cavity, paranasal sinuses and tympanic cavity of the koala. *Australian Veterinary Journal,* 91(9), 353–64.

Johnson, R.N., O'Meally, D., Chen, Z., Etherington, G.J., Ho, S.Y.W., Nash, W.J. et al. (2018). Adaptation and conservation insights from the koala genome. *Nature Genetics,* 50(8), 1102–11.

Lee, A.K. and Carrick, F.N. (1989). *Phascolarctidae. Fauna of Australia 1b: Mammalia, Volume 31.* Australian Government Printing Service, Canberra.

Mason, M.J. (2016). Structure and function of the mammalian middle ear. I: Large middle ears in small desert mammals. *Journal of Anatomy,* 228(2), 284–99.

Schmid, L.M., Schmid, K.L. and Brown, B. (1991). Behavioral determination of visual function in the koala (*Phascolarctos cinereus*). *Wildlife Research,* 18(3), 367–74.

Smith, M. (1980). Behavior of the koala, *Phascolarctos cinereus* (Goldfuss), in captivity 4. Scent-marking. *Australian Wildlife Research,* 7(1), 35–40.

Taylor, J., Brown, G., De Miguel, C., Henneberg, M. and Rühli, F.J. (2006). MR imaging of brain morphology, vascularisation and encephalization in the koala. *Australian Mammalogy,* 28(2), 243–47.

Toyota, M., Spencer, D., Sawai-Toyota, S., Jiaqi, W., Zhang, T., Koo, A.J., Howe, G.A. and Gilroy, S. (2018). Glutamate triggers long-distance, calcium-based plant defense signaling. *Science,* 361(6407), 1112–15.

Troughton, E. (1947). *Furred Animals of Australia.* Angus & Robertson, Sydney, 135–36.

V – Everything Changes

This reconstruction is based on publicly documented accounts by Indigenous people along the east coast of Australia as a means of reimagining a historical perspective history that has been excluded, ignored, misrepresented or erased from the historical record. Specific references used for this include the Dharawal calendar for the Port Jackson area at http://www.bom.gov.au/iwk/calendars/dharawal. shtml#marraigang, Illawarra cultural history from https://www.environment. nsw.gov.au/resources/cultureheritage/illawarraAboriginalHistoryPoster.pdf and 'The Message: The Story from the Shore' (2020) by Alison Page and the Bana Yirriji Artist group, which can be viewed at https://www.nma.gov.au/exhibitions/endeavour-voyage. Further references on this approach to Indigenous history and first contact accounts from the Twofold Bay area can be found in the story 'Spirit Brother' in Dooley G. and Clode, D. (eds) (2019) *The First Wave: Exploring Early Coastal Contact History in Australia*, Wakefield Press, Adelaide.

14 – Koalas Far and Wide

Briscoe, N.J., Krockenberger, A., Handasyde K.A. and Kearney M.R. (2015). Bergmann meets Scholander: geographical variation in body size and insulation in the koala is related to climate. *Journal of Biogeography*, 42(4), 791–802.

Cristescu, R.H., Foley, E., Markula, A., Jackson, G., Jones, D. and Frere, C. (2015). Accuracy and efficiency of detection dogs: a powerful new tool for koala conservation and management. *Scientific Reports*, 5, 8349.

Eyre, J.E. (1845). *Journals of Expeditions of Discovery into Central Australia*. T. and W. Boone, London.

Fuller, K. (2013). Could there be as few as 50 koalas left in the Pillaga? *ABC Local*, 15 December, https://www.abc.net.au/local/stories/2014/12/15/4148934. htm.

Johnson, R.N., O'Meally, D., Chen, Z., Etherington, G.J., Ho, S.Y.W., Nash, W. J. et al. (2018). Adaptation and conservation insights from the koala genome. *Nature Genetics*, 50(8), 1102–11.

Kelly, F. (2020). Koalas sighted once again in the Pilliga Forest of north-west NSW, *Radio National Breakfast*, 24 August, https://www.abc.net.au/radionational/programs/breakfast/koalas-sighted-once-again-in-the-pilliga-forest/12588396.

McDowell, M.C., Prideaux, G.J., Walshe, K., Bertuch, F. and Jacobsen G.E. (2015). Re-evaluating the Late Quaternary fossil mammal assemblage of Seton Rockshelter, Kangaroo Island, South Australia, including the evidence for late-surviving megafauna. *Journal of Quaternary Science*, 30, 355–64.

Neaves, L.E., Frankham, G.J., Dennison, S., FitzGibbon, S., Flannagan, C., Gillett, A. et al. (2016). Phylogeography of the koala (*Phascolarctos cinereus*) and harmonising data to inform conservation. *PLOS One*, 11(9), e0162207.

Nunn, P. (2018). *The Edge of Memory: Ancient Stories, Oral Tradition and the Post-Glacial World*. Bloomsbury Sigma, London.

Price, G.J. (2008). Is the modern koala (*Phascolarctos cinereus*) a derived dwarf of a Pleistocene giant? Implications for testing megafauna extinction hypotheses. *Quaternary Science Reviews*, 27, 2516–21.

Saltré, F., Chadoeuf, J., Peters, K.J., McDowell, M.C., Friedrich, T., Timmermann, A. et al. (2019). Climate-human interaction associated with southeast Australian megafauna extinction patterns. *Nature Communications*, 10(1), doi:10.1038/s41467-019-13277-0.

15 – A New Arrival

Notes: I have used the current preferred names of Indigenous nations and languages but recognise that these change over time. The Dreaming stories described are all 'open' stories told to children and strangers, as published (often without sources) by Mountford and Brough Smyth, Robinson (told by initiated Ngumbarr man James McGrath), Reed, Smith (in consultation with Gumbaynggirr Elders) and Walsh. To the best of my knowledge, no sacred or restricted knowledge has been shared in this chapter. Some of these stories are shared by Indigenous artist Michael Connolly (Munda gutta Kulliwari) at https://www.kullillaart.com.au/dreamtime-stories/. The jimbirn headband was collected from south-eastern Victoria in 1865 (Item X1574, Museums Victoria collections https://collections.museumsvictoria.com.au/items/158702). The other artefact representing koalas is a possum skin cloak made by Indigenous creator Phoebe Nicholson in 2013.

Brough Smyth, R. (1876). *The Aborigines of Victoria*, vol. 1. John Currey, O'Neil, Melbourne.

Cahir, F., Schlagloth, R. and Clark, I.D. (2020). The historic importance of the koala in Aboriginal society in New South Wales, Australia: an exploration of the archival record. *ab-Original*, 3(2), 172–91.

Cahir, F., Schlagloth, R. and Clark, I.D. (2021). The importance of the koala in Aboriginal society in nineteenth-century Queensland (Australia): a reconsideration of the archival record. *Anthrozoös*, doi:10.1080/08927936.2021.1963544.

Davies, N., Gramotnev, G., Seabrook, L., Bradley, A., Baxter, G., Rhodes, J. et al. (2013). Movement patterns of an arboreal marsupial at the edge of its range: a case study of the koala. *Movement Ecology*, 1(1), 8.

Dibden, J. (2011). Drawing in the land rock-art in the Upper Nepean, Sydney Basin, New South Wales. Unpublished PhD thesis, Australian National University, Canberra.

Govett, W. (1836). Sketches of New South Wales XIV. *Saturday Magazine*, 31 December, 8–9(288), 249–50.

Department of Environment and Conservation NSW. (2005). Early contact map: Why the whale spouts, the starfish is ragged, and the native bear has strong arms, in *A History of the Aboriginal People of the Illawarra Region*, https://www.environment.nsw.gov.au/resources/cultureheritage/illawarraAboriginalHistoryPoster.pdf

Campbell, A. and Vanderwal, R. (eds) (1999). *John Bulmer's Recollections of Victorian Aboriginal Life*. Museums Victoria Publishing, Melbourne, 62.

Johnson, C.N. (2016). Fire, people and ecosystem change in Pleistocene Australia. *Australian Journal of Botany*, 64(8), 643–51.

McDonald, J. (2008). Dreamtime superhighway (TA27): Sydney Basin rock art and prehistoric information exchange. ANU Press, Canberra.

Roberts, A. and Mountford, C.P. (1973). *The Dreamtime Book*. Rigby, Adelaide.

Munks, S.A., Corkrey, R. and Foley W.J. (1996). Characteristics of arboreal marsupial habitat in the semi-arid woodlands of northern Queensland. *Wildlife Research*, 23, 185–95.

Nunn, P.D. and Reid, N.J. (2016). Aboriginal memories of inundation of the Australian coast dating from more than 7000 years ago. *Australian Geographer*, 47(1), 11–47.

Parris, H.S. (1948). Koalas on the Lower Goulburn. *Victorian Naturalist*, 64, 192–93.

Phillips, B. (1990). *Koalas: The Little Australians We'd All Hate to Lose*. Australian National Parks and Wildlife Service, Canberra.

Reed, A.W. (1965). *Myths and Legends of Australia*. A.H. and A.W. Reed, Sydney.

Robinson, R. (1965). *The Man Who Sold His Dreaming*. Currawong Publishing, Sydney.

Schlagloth, R., Cahir F. and Clark I. (2018). The importance of the koala in Aboriginal society in nineteenth-century Victoria (Australia): A reconsideration of the archival record. *Anthrozoös*, 31(4), 433–41.

Smith, S., Smith, N., Daley, L., Wright, S. and Hodge, P. (2021). Creation, destruction, COVID-19: Heeding the call of country, bringing things into balance. *Geographical Research*, 59(2), 160–68.

Sullivan, B.J., Baxter, G.S. and Lisle, A.T. (2003). Low-density koala (*Phascolarctos cinereus*) populations in the mulgalands of south-west Queensland. III. Broad-scale patterns of habitat use. *Wildlife Research*, 30, 583–91.

Taçon, P.S.C., South, B. and Boree Hooper, S. (2003). Depicting cross-cultural interaction: figurative designs in wood, earth and stone from south-east Australia. *Archaeology in Oceania*, 38, 89–101.

Walsh, G.L. (1985). *Didane the Koala*. University of Queensland Press, St Lucia.

16 – The English Annexation

Notes: The first major smallpox epidemic broke out among Indigenous communities fifteen months after the arrival of the First Fleet and has variously (and implausibly) been blamed on the French and Makassar sources. It is known that the First Fleet surgeon John White carried with him a glass jar of the smallpox virus intended to inoculate children, although whether this was infectious is unknown. The location of the earliest sighting of koalas in New South Wales is often given as the Blue Mountains, but this is incorrect. Koalas are mentioned in some of the Bullock catalogues (1810, 1811) but missing from others (1812, 1817).

Anon. (1803). [First account of live koala], *The Sydney Gazette and New South Wales Advertiser*, Sunday 21 August, 3.

Bullock, W. (1810). *A Companion to Mr. Bullock's London Museum and Pantherion*. William Bullock, London.

Clode, D. (2007). *Voyages to the South Seas*. Ligature, Balmain.

de Blainville, M.H. (1816). *Prodome d'une nouvelle distribution systématique de règne animal*, Bulletin de sciences, Impremiere de Plassan, Paris, 105–24.

Flynn, M. and Sturgess, G. (2015). New evidence on Arthur Phillip's first landing place 26 January 1788, *Royal Australian Historical Society*, https://www.rahs.org.au/wp-content/uploads/2015/12/New-Evidence-Online-Version1.pdf.

Geoffroy Saint-Hilaire, E. (1826). Koala. In Audoiin, I. et al., *Dictionnaire Classique d'Histoire Naturelle*, vol. 9, Rey et Graviers, Paris, 133–34.

Govett W.R. (1836). Sketches of New South Wales, no. XIV, *The Saturday Magazine*, December, 8–9 (288), 249–50.

Home, E. (1832). An account of some peculiarities in the anatomical structure of the wombat, with observations on the female organs of generation. *Abstracts of the Papers Printed in the Philosophical Transactions of the Royal Society of London*, 1, 310–11.

Lunney, D., Close, R., Bryant, J., Crowther, M.S., Shannon, I., Madden, K. and Ward, S. (2010). Campbelltown's koalas: their place in the natural history of Sydney. In Lunney, D., Hutchings, P. and Hochuli, D. (eds), *The Natural History of Sydney*, Royal Zoological Society of NSW, Mosman, 319–25.

Moyal, A. (2008). *Koala: A Historical Biography*, CSIRO Publishing, Canberra.

Perry, G. (1810). Kaolo, or New Holland sloth, *Arcana, or The Museum of Natural History*. George Smeeton, London. 65–9, Plate 17.

Petit, R.W. (2010). *Perry's Arcana: A Facsimile Edition*. Temple University Press, Philadelphia.

Watkin, T. and Walker, J. (1793). A map of the hitherto explored country contiguous to Port Jackson: Lain down from actual survey, J. Walker, G. Nicol, London. Retrieved from State Library of New South Wales, https://search.sl.nsw.gov.au/permalink/f/lg5tom/SLNSW_ALMA2196069000002626.

17 – War and Guns

Notes: Images of koalas as mascots in the World War I can be found in *The Queenslander*, 19 December 2014, 23, https://trove.nla.gov.au/newspaper/page/2503514 and the Australian War Memorial archive available at https://www.awm.gov.au (see images C194709, C1192404, C1230886, C214118 and C1194203).

Anon. (1921). The koala. *The Catholic Press* (Sydney), 15 December, 43, http://nla.gov.au/nla.news-article106256258.

Anon. (1927). Protection for the koala, *Daily Mercury* (Mackay), 23 July, 15, http://nla.gov.au/nla.news-article173770794.

Eberhard, I.H. (1972). *Ecology of the Koala (Phascolarctos cinereus, Goldfuss)*

on Flinders Chase, Kangaroo Island. Unpublished PhD thesis, University of
Adelaide.

Gelder, K. and Weaver, R. (2020) *The Colonial Kangaroo Hunt*. Miegunyah
Press, Melbourne.

Howlett, N.L. (1979). The bear you couldn't buy: shooting koalas in Queensland:
1927. *Bowyang* 1(2), 9–24.

Phillips, B. (1990). *Koalas: The Little Australians We'd All Hate to Lose*.
Australian National Parks and Wildlife Service, Canberra.

Taylor, S. (2022). Rose and Morris Michtom: The couple behind ideal toy company,
Collectors Quests, https://collectorsquests.com/rose-morris-michtom-
ideal-toy-company/.

Troughton, E.L.G. (1938). Australian mammals: their past and future. *Journal of
Mammalogy*, 19(4), 401–11.

US National Parks Service. (2021). The story of the teddy bear, *Theodore
Roosevelt Birthplace, National Historic Site New York*, https://www.nps.
gov/thrb/learn/historyculture/storyofteddybear.htm.

Whyte, J. (2010). Vintage Koala Toys, https://www.ubear.com.au/vintage-
koala-toys.

18 – Saving the Koala

Notes: Local information on the extinction of koalas in the south-west of South
Australia are from Tim Crofts, based on significant fire damage in 1933–34,
1938–39 and 1943–44: https://www.cfs.sa.gov.au/about-cfs/history-of-the-cfs/
bushfire-history.

Anon. (1933). Blinky Bill [review]. *Northern Standard* (Darwin), 1 December,
6, http://nla.gov.au/nla.news-article48063903.

Anon. (1937). Minister supports koala bear preservation. *News* (Adelaide),
12 January, 3, http://nla.gov.au/nla.news-article131400343.

Anon. (1939). Koala bears thriving at Flinders Chase. *The Advertiser* (Adelaide),
21 February, 18, http://nla.gov.au/nla.news-article35582690.

Anon. (1944). Condition of koalas: Quail Island sanctuary. *The Age* (Melbourne),
25 January, 3, http://nla.gov.au/nla.news-article206780585.

Anon. (1948) Gums for Kangaroo Island bears. *The Advertiser* (Adelaide),
11 December, 6, http://nla.gov.au/nla.news-article43795201.

Anon. (1948). Koalas for Kangaroo Island. *Barrier Miner* (Broken Hill),
12 November, 5, http://nla.gov.au/nla.news-article48577652.

Anon. (1949). Koala bears. *The Advertiser* (Adelaide), 1 January, 2, http://nla.
gov.au/nla.news-article43798010.

Canfield, P.J. (1993). Disease and mortality in Australasian marsupials held at
London Zoo, 1872–1972. *Journal of Zoo and Wildlife Medicine*, 24, 158–67.

Clode, D. (2010). *A Future in Flames*. Ligature, Balmain.

Clode, D. (2014). Seeing the wood for the trees. *Australian Book Review*,
November, 366, 40–50.

Forbes, W.A. (1881). On some points in the anatomy of the koala (*Phascolarctos cinereus*). *Proceedings of the Zoological Society of London*, 49(1), 180–95.

Igulden, W. (1943). Quail Island koalas. *The Argus* (Melbourne), 19 November, 5, http://nla.gov.au/nla.news-article11801880.

Martin, R. and Handasyde, K. (1999). *The Koala: Natural History, Conservation and Management*. UNSW Press, Sydney.

McVitty, W. (1988). *Dorothy Wall, the Creator of Blinky Bill: Her Life and Work – A Biography*. Angus and Robertson, North Ryde.

Menkhorst, P. (2008). Hunted, marooned, re-introduced, contracepted: a history of koala management in Victoria. In D. Lunney, A. Munn and W. Meikle (eds), *Too Close for Comfort: Contentious Issues in Human-Wildlife Encounters*, Royal Zoological Society of NSW, Mosman, 73–92.

VI – Future Tense

This reconstruction is based on the natural cycle of burning in mountain ash forests, which predominantly regenerate only after rare, extreme fires.

19 – Sex, Disease and Genetic Diversity

Notes: The Bramble Cay melomys was declared extinct on 22 February 2019, Threatened Species Scientific Committee Listing Advice *Melomys rubida*, Bramble Cay melomys, http://www.environment.gov.au/biodiversity/threatened/species/pubs/64477-listing-advice-22022019.pdf.

Australian Koala Foundation (2020). Distribution, https://www.savethekoala.com/about-koalas/distribution/.

Berlage, E. (2020). Relaxed and healthy: rare spotting on koala pair in Grampians. *The Wimmera Mail-Times*, 22 January, https://www.mailtimes.com.au/story/6591387/relaxed-and-healthy-rare-spotting-of-koala-pair-in-grampians.

Breed, B. and Ford, F. (2007). *Native Mice and Rodents*. CSIRO Publishing, Collingwood.

Cristescu, R., Cahill, V., Sherwin, W.B., Handasyde, K., Carlyon, K., Whisson, D. et al. (2009). Inbreeding and testicular abnormalities in a bottlenecked population of koalas (*Phascolarctos cinereus*). *Wildlife Research*, 36(4), 299–308.

Degiorgio, M., Jakobsson, M. and Rosenberg, N.A. (2009). Explaining worldwide patterns of human genetic variation using a coalescent-based serial founder model of migration outward from Africa. *Proceedings of the National Academy of Sciences*, 106(38), 16057–62.

Fabijan, J., Caraguel, C., Jelocnik, M., Polkinghorne, A., Boardman, W.S.J., Nishimoto, E. et al. (2019). *Chlamydia pecorum* prevalence in South Australian koala (*Phascolarctos cinereus*) populations: Identification and modelling of a population free from infection. *Scientific Reports*, 9(1), 6261.

Greenwood, A.D., Ishida, Y., O'Brien, S.P., Roca, A.L. and Eiden, M.V. (2018). Transmission, evolution, and endogenization: lessons learned from recent retroviral invasions. *Microbiology and Molecular Biology Reviews*, 82(1), e00044–00017.

Jelocnik, M. and Polkinghorne, A. (2017). Chlamydia pecorum: successful pathogen of koalas or Australian livestock? *Microbiology Australia*, 38(3), 101–05.

Johnson, R.N., O'Meally, D., Chen, Z., Etherington, G.J., Ho, S.Y.W., Nash, W.J. et al. (2018). Adaptation and conservation insights from the koala genome. *Nature Genetics*, 50(8), 1102–11.

Kean, Z.(2019). Koala retrovirus leads scientists to discover 'second immune system'. *ABC Science*, 12 October, https://www.abc.net.au/news/science/2019-10-12/koala-retro-virus-shed-light-on-second-immune-system/11583782.

Kjeldsen, S.R., Raadsma, H.W., Leigh, K.A., Tobey, J.R., Phalen, D., Krockenberger, A. et al. (2019). Genomic comparisons reveal biogeographic and anthropogenic impacts in the koala (*Phascolarctos cinereus*): a dietary-specialist species distributed across heterogeneous environments. *Heredity*, 122(5), 525–44.

McMichael, L., Smith, C., Gordon, A., Agnihotri, K., Meers, J. and Oakey, J. (2019). A novel Australian flying-fox retrovirus shares an evolutionary ancestor with Koala, Gibbon and Melomys gamma-retroviruses. *Virus Genes*, 55(3), 421–24.

Pike, B.L., Saylors, K.E., Fair, J.N., LeBreton, M., Tamoufe, U., Djoko, C.F. et al. (2010). The origin and prevention of pandemics. *Clinical Infectious Diseases*, 50(12), 1636–40.

Robins, B. (1978). 'Cleland Koalas: a report on the koalas at Cleland Park', unpublished report courtesy of C.B. Daniels.

Teixeira, J.C. and Huber, C.D. (2021). The inflated significance of neutral genetic diversity in conservation genetics. *Proceedings of the National Academy of Sciences*, 118(10), e2015096118.

Thorpe, N., Willis, O. and Smith, C. (2021). 'Devil, devil': the sickness that changed Australia. *ABC News,* 18 August, https://www.abc.net.au/news/health/2021-06-07/patient-zero-smallpox-outbreak-of-1789/100174988.

Troughton, E. (1962). *Furred Animals of Australia*, Angus and Robertson: Sydney, 136.

Tsangaras, K., Avila-Arcos, M.C., Ishida, Y., Helgen, K.M., Roca, A L. and Greenwood, A.D. (2012). Historically low mitochondrial DNA diversity in koalas (Phascolarctos cinereus). *BMC Genetics*, 13(1), 92.

Yu, T., Koppetsch, B.S., Pagliarani, S., Johnston, S., Silverstein, N.J., Luban, J. et al. (2019). The piRNA response to retroviral invasion of the koala genome. *Cell*, 179(3), 632–43.e612.

20 – Expansion and Retreat

Notes: Legislation governing koalas in Australia includes the *Wildlife Protection (Regulation of Exports and Imports) Act 1982*, https://www.legislation.gov.au/Series/C2004A02708, which was replaced by the *Environment Protection and*

Biodiversity Conservation Amendment (Wildlife Protection) Act 2001, https://www.legislation.gov.au/Details/C2004A00849. State laws regulate the keeping of native animals as pets, while the transfer of koalas overseas is covered by the Conditions for the Overseas Transfer of Koalas regulations, Department of Environment, Water, Heritage and the Arts (2009), https://www.awe.gov.au/sites/default/files/documents/koala-export-conditions.pdf.

Adams-Hosking, C., Grantham, H.S., Rhodes, J.R., McAlpine, C. and Moss, P.T. (2011). Modelling climate-change-induced shifts in the distribution of the koala. *Wildlife Research*, 38(2), 122–30.

Anon. (2015). Starving koalas secretly culled at Cape Otway, 'overpopulation issues' blamed for ill health, *ABC News*, 4 March, https://www.abc.net.au/news/2015-03-04/starving-koalas-secretly-culled-at-cape-otway/6278768.

Booth, T.H. (1998). Biodiversity and conservation of Australian forests, *Commonwealth Forestry Review*, 77, 39–46.

Branigin, W. (1983). Forget the piddling koalas. *The Washington Post*, 22 May, https://www.washingtonpost.com/archive/politics/1983/05/22/forget-the-piddling-koalas/b038a1d8-1401-45fa-afe2-432f17b2952a/.

Ellis, W., Melzer, A., Clifton, I. and Carrick, F. (2010). Climate change and the koala *Phascolarctos cinereus*: water and energy. *Australian Zoologist*, 35(2), 369–77.

Fowles, S. (2015). Cape Otway koala cull to return with wildlife officers to assess 400 koalas. *Geelong Advertiser*, 14 September, https://www.geelongadvertiser.com.au/news/geelong/cape-otway-koala-cull-to-return-with-wildlife-officers-to-assess-400-koalas/news-story/101d08b02e9fbfafbd7e-a4e9a72ae7af.

Kingsford, R.T., Dunn, H., Love, J., Nevill, J., Stein, J. and Tait, J. (2005). *Protecting Australia's rivers, wetlands and estuaries of high conservation value*. Department of Environment and Heritage Australia, Canberra, Product Number PR050823, https://www.awe.gov.au/sites/default/files/documents/protecting-rivers.pdf

Kyodo News. (2014). Koalas' prodigious costs, graying population bedevil nation's zookeepers. *The Japan Times*, 22 September.

Montreal Process Implementation Group for Australia and National Forest Inventory Steering Committee. (2018). *Australia's State of the Forests Report 2018*, ABARES, Canberra, December, https://www.agriculture.gov.au/abares/forestsaustralia/forest-data-maps-and-tools/native-forest-data-visualisation.

Schlagloth, R., Santamaria, F., Melzer, A., Keatley, M.R. and Houston, W. (2021). Vehicle collisions and dog attacks on Victorian koalas as evidenced by a retrospective analysis of sightings and admission records 1997–2011. *Australian Zoologist*, https://doi.org/10.7882/AZ.2021.030.

Shabani, F., Ahmadi, M., Peters, K.J., Haberle, S., Champreux, A., Saltre, F. and Bradshaw, C.J.A. (2019). Climate-driven shifts in the distribution of koala-browse species from the Last Interglacial to the near future. *Ecography*, 42(9), 1587–99.

Standing Council on Environment and Water. (2011). *Australia's Native Vegetation Framework*, Department of Sustainability, Environment, Water, Population and Communities, Canberra, 20–21, https://www.awe.gov.au/sites/default/files/documents/native-vegetation-framework.pdf.

21 – Under Fire

Batsakis, A. and Mountain, W. (2020). Click through the tragic stories of 119 species still struggling after Black Summer in this interactive (and how to help), *The Conversation*, 14 July, https://theconversation.com/click-through-the-tragic-stories-of-119-species-still-struggling-after-black-summer-in-this-interactive-and-how-to-help-131025.

Briscoe, N.J., Handasyde, K.A., Griffiths, S.R., Porter, W.P., Krockenberger, A. and Kearney, M.R. (2014). Tree-hugging koalas demonstrate a novel thermoregulatory mechanism for arboreal mammals. *Biology Letters*, 10(6), 5. doi:10.1098/rsbl.2014.0235.

Chen, K., and McAneney, J. (2004). Quantifying bushfire penetration into urban areas in Australia. *Geophysical Research Letters*, 31, 1–4.

Claughton, D. (2021). Animal rescue groups race to save Australia's bushfire-hit wildlife with nearly $200 million in donations. *ABC News*, 26 February, https://www.abc.net.au/news/2021-02-26/animal-rescuers-raise-$200-million-for-bushfire-hit-wildlife/13135592.

Clode, D. (2019). Living with fire and facing our fears. *The Conversation*, 6 December, https://theconversation.com/friday-essay-living-with-fire-and-facing-our-fears-128093.

Dunstan, E., Funnell, O., McLelland, J., Stoeckeler, F., Nishimoto, E., Mitchell, D. et al. (2021). An analysis of demographic and triage assessment findings in bushfire-affected koalas (*Phascolarctos cinereus*) on Kangaroo Island, South Australia, 2019–2020. *Animals*, 11(11), 3237.

Fuge, L. (2021). More megafires loom in Australia's future. *Cosmos*, 26 November, https://cosmosmagazine.com/earth/climate/megafires-in-australia-will-increase-new-study-shows/.

Gorton, S. (2020). More than 240 GoFundMe pages raise $3.8 million for Kangaroo Island bushfire recovery. *The Islander*, 6 February, https://www.theislanderonline.com.au/story/6615003/more-than-240-gofundme-pages-raise-38-million-for-kangaroo-island-bushfire-recovery/.

Morton, A. (2020). 'Dinosaur trees': firefighters save endangered Wollemi pines from NSW bushfires. *Guardian Australia*, 15 January, https://www.theguardian.com/australia-news/2020/jan/15/dinosaur-trees-firefighters-save-endangered-wollemi-pines-from-nsw-bushfires.

Smith, A.G., McAlpine, C.A., Rhodes, J.R., Lunney, D., Seabrook, L. and Baxter, G. (2013). Out on a limb: habitat use of a specialist folivore, the koala, at the edge of its range in a modified semi-arid landscape. *Landscape Ecology*, 28(3), 415–26.

22 – In a Perfect World

Australian Conservation Foundation. (2022). Aggravating extinction investigation: How the Australian government approves the destruction of threatened species habitat. Carlton, Victoria, www.acf.ofg.au/reports.

Department of Agriculture, Water and the Environment. (2022). *EPBC Act List of Threatened Fauna*, https://www.environment.gov.au/cgi-bin/sprat/public/publicthreatenedlist.pl#mammals_endangered (accessed 5 April 2022).

Pacheco, P., Mo, K., Dudley, N., Shapiro, A., Aguilar-Amuchastegui, N., Ling, P.Y., Anderson, C. and Marx, A. (2021). Deforestation fronts: Drivers and responses in a changing world. WWF, Gland, Switzerland, https://wwfint.awsassets.panda.org/downloads/deforestation_fronts___drivers_and_responses_in_a_changing_world___full_report_1.pdf.

Schlagloth, R., Golding, B., Kentish, B., McGinnis, G., Clark, I.D., Cadman, T., Cahir, F. and Santamaria, F. (2022). Koalas – Agents for Change: A case study from regional Victoria. *Journal of Sustainability Education*, 26 February.

Woinarski, J. and Burbidge, A.A. (2020). *Phascolarctos cinereus* (amended version of 2016 assessment). *The IUCN Red List of Threatened Species* 2020: e.T16892A166496779, https://dx.doi.org/10.2305/IUCN.UK.2020-1.RLTS.T16892A166496779.en. (accessed on 7 April 2022).

INDEX